机械行业英语——职场英语

主　编：刘亚丽　孟　皎　张宏杰
副主编：于晓丽　车君华　张雅茹　李　莉
　　　　刘连杰　付　敏
参　编：李　磊　于泽鑫　王　勇　张泽衡
　　　　徐西华　丁明辉　李培积　冀永帅

北京理工大学出版社
BEIJING INSTITUTE OF TECHNOLOGY PRESS

版权专有　侵权必究

图书在版编目（CIP）数据

机械行业英语：职场英语 / 刘亚丽，孟皎，张宏杰主编. —北京：北京理工大学出版社，2019.8

ISBN 978-7-5682-7398-5

Ⅰ. ①机⋯　Ⅱ. ①刘⋯ ②孟⋯ ③张⋯　Ⅲ. ①机械工业 – 英语 – 高等学校 – 教材　Ⅳ. ①TH

中国版本图书馆 CIP 数据核字（2019）第 174531 号

出版发行 / 北京理工大学出版社有限责任公司
社　　址 / 北京市海淀区中关村南大街 5 号
邮　　编 / 100081
电　　话 /（010）68914775（总编室）
　　　　　（010）82562903（教材售后服务热线）
　　　　　（010）68948351（其他图书服务热线）
网　　址 / http://www.bitpress.com.cn
经　　销 / 全国各地新华书店
印　　刷 / 北京国马印刷厂
开　　本 / 787 毫米 × 1092 毫米　1/16
印　　张 / 9.5　　　　　　　　　　　　　　　　　　责任编辑 / 武丽娟
字　　数 / 223 千字　　　　　　　　　　　　　　　　文案编辑 / 武丽娟
版　　次 / 2019 年 8 月第 1 版　2019 年 8 月第 1 次印刷　责任校对 / 刘亚男
定　　价 / 45.00 元　　　　　　　　　　　　　　　　责任印制 / 李志强

图书出现印装质量问题，请拨打售后服务热线，本社负责调换

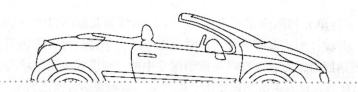 

前言 PREFACE

2015年，国务院颁布了《关于推进国际产能和装备制造合作的指导意见》，其中要求加快职业教育"走出去"步伐，支持"一带一路"倡议的软实力发展。作为世界通用语言，英语尤其是与专业紧密结合的行业英语教学符合国家发展战略要求。《中国制造2025》是中国制造迈向中高端的有利契机，产业转型升级对高端技术技能人才培养，包括员工的英语语言能力培养提出了更高要求。

《机械行业英语》系列教材立足于德国"工业4.0"和《中国制造2025》对高素质技术技能人才的国际化需求，紧密结合生产制造类岗位实际，以"职场英语+机械职业英语"的课程体系为基础，按照学生职业成长规律进行系统化设计，旨在训练学生在职场环境中的英语听说读写能力、在职业情境中准确使用英语进行交际的能力以及团队合作等职业素养。

《职场英语》是系列教材的第一部分，本书共8个单元，充分考虑到英语与专业、普通英语与职业英语的衔接，融会贯通。每单元内容编排丰富，涵盖了职业英语听说、读写、词汇及背景文化知识等多维度能力要求。本书在编写中力求达到以下特点：

1. 按照职业生涯发展顺序设计情境。从学生入学、选择专业、与企业签约等一系列双元制学生入学、双选会情境入手，逐步延伸到车间、企业，兼顾6S管理、安全、精益生产等职业素养，构建基于情境的英语教材，教材内容均来源于合作企业的岗位需求分析，与专业、与学生未来就业岗位密切相关。

每个单元分为课程准备、口语训练、读写拓展、交际技巧四个部分，围绕同一主题进行背景引入和词汇、口语训练，并结合企业实际案例进行读写拓展，循序渐进，逐步深入。在对话练习的基础上提炼总结交际技巧，帮助学生练习地道的口语表达。各情境编写框架如下：

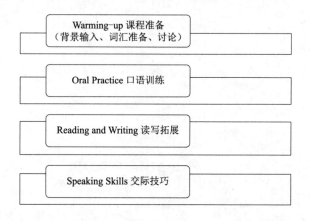

学生在情境中学习职场常用的英语口语表达，练习听说读写技巧，提升综合英语能力。本教材不追求英语语法的完整和体系，体现高职教育教学内容选择时"必需""够用"的原则。

2. 重视口语交际能力培养。每单元均围绕某一主题设计了以情境为基础的口语对话，应用丰富多样的口语惯用表达，并设计形式多样的练习供学生以小组为单位进行口语练习，做到即学即练。强调学生口语表达能力，力求改变以往英语课堂中的"哑巴式英语"状况。根据单元教学内容提炼总结口语技巧，重点讲解副词使用、连接词、委婉表达等地道的英语口语表达法，力争提升学生的口语交际能力。

3. 职业核心素养培养贯穿始终。教材中词法和句法知识的讲解练习以及延伸拓展练习旨在训练学生学词组句、查阅资料等自主学习能力，从而为终身学习打下坚实基础。行业知识、专业知识，如工具类词汇、设备说明书中常见的短语、语法、用法的渗透，可为进一步学习《机械职业英语》做好铺垫。教材中还设计了讨论等开放性问题，鼓励学生积极思考、大胆创新。多种形式的小组合作任务，要求学生开展合作学习，提升表达能力和团队合作意识。

本书可作为职业院校的机械制造与自动化类专业、数控技术、机电类专业的英语教材使用，也适合从事机械行业的社会人员自学使用。

本书由刘亚丽、孟皎、张宏杰担任主编。具体编写分工如下：刘亚丽负责编写6~8单元，1~5单元由孟皎和张宏杰共同编写，于晓丽负责编写本书交际技巧和附录部分，刘亚丽负责统稿和定稿。本书由车君华、张雅茹策划、构思并设计了教材结构和本课程的整体知识结构。李莉、刘连杰、付敏参与了资料的收集整理工作。参与编写的还有合作企业的同仁和培训师们，他们为全书的情境设计提供了诸多素材、建议和意见。

本书在编写过程中参考了大量文献资料，在此谨向原著者表示诚挚的谢意。由于编者水平有限，加之编写时间仓促，书中难免有不妥之处，敬请广大读者批评指正。

编　者

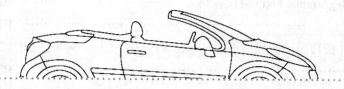

目录
CONTENTS

Unit 1　Orientation Day 新生入学 ········· 001
  Warming-up 课程准备 ········· 001
    Mission Objective 任务目标 ········· 001
    Discussion 讨论 ········· 001
    Database 知识库 ········· 001
  Oral Practice 口语训练 ········· 002
    Dialogue 1　Meeting the New Student 遇见新同学 ········· 003
    Dialogue 2　Visiting the Campus 参观校园 ········· 004
    Dialogue 3　Asking the Way 问路 ········· 005
  Reading and Writing 读写拓展 ········· 008
    Reading 阅读 ········· 008
    Writing 写作 ········· 009
      Leave/Absence Request Form 请假申请 ········· 009
      Request for Leave 请假条 ········· 009
  Speaking Skills 交际技巧 ········· 011
    妙词巧句 ········· 011

Unit 2　Major Introduction 专业介绍 ········· 013
  Warming-up 课程准备 ········· 013
    Mission Objective 任务目标 ········· 013
    Discussion 讨论 ········· 013
    Database 知识库 ········· 013
  Oral Practice 口语训练 ········· 014
    Dialogue 1　Consultation through Telephone 电话咨询 ········· 015
    Dialogue 2　Industrial Process Automation Technology
                  工业过程自动化技术专业 ········· 017
    Dialogue 3　CNC Technology 数控技术专业 ········· 019
  Reading and Writing 读写拓展 ········· 021

Reading 阅读 ··· 021
Writing 写作 ·· 023
    Conference Registration Form 会议签到表 ····························· 023
    Minutes 会议记录 ·· 023
Speaking Skills 交际技巧 ··· 026
    巧用程度副词 ··· 026

Unit 3　Job Fair 双选会 ·· 028

Warming-up 课程准备 ··· 028
    Mission Objective 任务目标 ·· 028
    Discussion 讨论 ·· 028
    Database 知识库 ·· 028
Oral Practice 口语训练 ··· 030
    Dialogue 1　Consulting the Teacher about the Job Fair 向老师咨询双选会 ··· 030
    Dialogue 2　Discussing the Future Job with Classmates
               与同学讨论未来的工作 ····························· 032
    Dialogue 3　During the Interview 进行面试 ······························ 033
Reading and Writing 读写拓展 ··· 035
    Reading 阅读 ·· 035
    Writing 写作 ·· 037
        Employee Entry Registration Form 员工入职登记表 ············· 037
Speaking Skills 交际技巧 ··· 039
    巧用连接词 ··· 039

Unit 4　Working Environment 工作环境 ······································· 042

Warming-up 课程准备 ··· 042
    Mission Objective 任务目标 ·· 042
    Discussion 讨论 ·· 042
    Database 知识库 ·· 042
Oral Practice 口语训练 ··· 044
    Dialogue 1　Workshop of Manual Machining 手动加工车间 ········· 045
    Dialogue 2　Workshop of Machine-tool Processing 机床加工车间 ········ 046
    Dialogue 3　Knowing the Factory 认识厂区 ····························· 048
Reading and Writing 读写拓展 ··· 049
    Reading 阅读 ·· 049
    Writing 写作 ·· 051
        Training Application Form 培训申请单 ································ 051
        Application Letter 申请信 ··· 051

Speaking Skills 交际技巧 ·········· 053
　话语填充 ·········· 053

Unit 5　Health and Safety 健康与安全 ·········· 055
Warming-up 课程准备 ·········· 055
　Mission Objective 任务目标 ·········· 055
　Discussion 讨论 ·········· 055
　Database 知识库 ·········· 055
Oral Practice 口语训练 ·········· 056
　Dialogue 1　Using Fire Extinguishers 使用灭火器 ·········· 057
　Dialogue 2　An Accident in the Office 办公室事故 ·········· 059
　Dialogue 3　Safety Instructions in Workshop 车间安全指导 ·········· 060
Reading and Writing 读写拓展 ·········· 062
　Reading 阅读 ·········· 062
　Writing 写作 ·········· 064
　　Non-conformance Report 不合格品报告 ·········· 064
　　E-mail 电子邮件 ·········· 065
Speaking Skills 交际技巧 ·········· 068
　巧用替换 ·········· 068

Unit 6　6S Management　6S 管理 ·········· 070
Warming-up 课程准备 ·········· 070
　Mission Objective 任务目标 ·········· 070
　Discussion 讨论 ·········· 070
　Database 知识库 ·········· 071
Oral Practice 口语训练 ·········· 072
　Dialogue 1　Importance of 6S Management 6S 管理的重要性 ·········· 072
　Dialogue 2　Explanation of 6S Management 6S 管理的含义 ·········· 074
Reading and Writing 读写拓展 ·········· 077
　Reading 阅读 ·········· 077
　Writing 写作 ·········· 078
　　Rationalization Proposal Form 合理化建议表 ·········· 078
　　Letter of Recommendation 建议信 ·········· 079
Speaking Skills 交际技巧 ·········· 081
　委婉表达 ·········· 081

Unit 7　Office English 办公英语 ·········· 084
Warming-up 课程准备 ·········· 084
　Mission Objective 任务目标 ·········· 084

Discussion 讨论 ·············· 084
Database 知识库 ·············· 084
Oral Practice 口语训练 ·············· 086
　Dialogue 1　Arriving at the Office 初到办公室 ·············· 086
　Dialogue 2　Getting to know the Equipments and Duties
　　　　　　　了解办公设备和工作职责 ·············· 087
　Dialogue 3　Face-to-Face Regular Weekly 每周面谈 ·············· 090
Reading and Writing 读写拓展 ·············· 091
　Reading 阅读 ·············· 091
　Writing 写作 ·············· 092
　　Customer Satisfaction Survey 顾客满意度调查 ·············· 092
　　Letter of Complaint 投诉信 ·············· 094
Speaking Skills 交际技巧 ·············· 095
　对话破冰与结尾 ·············· 095

Unit 8　Lean Production 精益生产 ·············· 098

Warming-up 课程准备 ·············· 098
　Mission Objective 任务目标 ·············· 098
　Discussion 讨论 ·············· 098
　Database 知识库 ·············· 098
Oral Practice 口语训练 ·············· 100
　Dialogue 1　Standard Work 标准化工作 ·············· 100
　Dialogue 2　Eliminating Wastes 消除浪费 ·············· 101
　Dialogue 3　Team Meetings 班组会 ·············· 103
Reading and Writing 读写拓展 ·············· 104
　Reading 阅读 ·············· 104
　Writing 写作 ·············· 106
　　Summary after Production 产后总结 ·············· 106
　　Report 报告 ·············· 106
Speaking Skills 交际技巧 ·············· 108
　定语从句 ·············· 108

参考译文 ·············· 111
参考答案 ·············· 120
常用口语句型 ·············· 133
常用前缀后缀 ·············· 137
数词的译法 ·············· 139
科技英语中的被动语态及译法 ·············· 141

Unit 1

Orientation Day 新生入学

❄ Warming-up 课程准备

Mission Objective 任务目标

1. 掌握工程学、全日制学生、系部、专业、理论、实践、录取、学制、双语教学、毕业、教学楼等学校场所、工业过程自动化技术专业、数控技术专业、学徒等的英文表达，了解常见的词语前缀。
2. 会用英语问候、介绍、致谢、谈论个人信息、问路。
3. 会撰写英文请假条。

Discussion 讨论

1. What is your first impression of this college?
2. How is your first day on the campus?

Words and phrases for assistance:

人：freshman, upperclassman, senior, warm, friendly, helpful

事：sign my name, take photos, hand in the Certificate of Admission, visit the campus

地点：campus, playground, teaching building, dorm, office, canteen, library, supermarket

Database 知识库

Aachen University of Applied Sciences

The Aachen University of Applied Sciences, founded in 1971, is one of Germany's leading universities of applied technology. It has regional natural science teaching and research centers, research and development centers and technology transformation centers. Its engineering specialty ranks second among similar universities in Germany (University of Applied Technology, the same below). Not only does it have abundant research funds, but it also has close academic and scientific relations with European largest research center, which has more than 4,000 scientists and engineers from all over the world, where students can do internships and graduation theses.

The university has 12 departments and nearly 10,000 full-time students. About 2,000 international students from all over the world study here, accounting for about 21% of the total number of students. There are 250 professors and 450 faculty members. There are more than 30 majors in 12 departments. The courses emphasize the combination of theory and practice, and the subject settings are adapted to the development of applied science and technology. Due to its teaching and research strength, the school has received substantial financial support from the federal government, the state government, the European Union, and industrial and commercial enterprises from all walks of life, attracting more and more students to choose to enroll in the university.

The university is developing towards internationalization. At present, it has established partnerships with more than 130 universities in the United States, China, Canada, Britain, France and Australia , including mutual recognition of academic qualifications and conversion credits. Students who enter the university can also apply to transfer to partner schools in other countries. The university offers the most international degree programs and the best results in similar universities in Germany.

The school offers bilingual courses in English and German for international students, with a four-year academic system and a bachelor's degree after graduation. These courses for international students are taught in English, and bilingual teaching is used in professional study, so that international students can study directly in Germany without a German-speaking basis.

Activity 1: Fill in the blanks according to the passage. 根据短文填空。

1. 应用技术大学_____
2. 技术转化中心_____
3. 工程学_____
4. 实习_____
5. 全日制学生_____
6. 系部_____
7. 专业_____
8. 教职工_____
9. 理论_____
10. 实践_____
11. 录取_____
12. 双语教学_____
13. 学制_____
14. 学士学位_____
15. 毕业_____

Oral Practice 口语训练

人物涉及 (Characters):

Z: Zhang Yang (volunteer of Jinan Vocational College, majoring in CNC (Computerized Numerical Control) Technology, signed apprentice of FESTO) 济南职业学院义工，数控技术专业学生，FESTO 签约学徒

W: Wang Nan (freshman majoring in Industrial Process Automation Technology) 工业过程自动化技术专业新生

情景解说 (Situations)：

Wang Nan(W) 是工业过程自动化技术专业新生，入学第一天学校义工 Zhang Yang(Z) 向他介绍学院信息，引导他到宿舍并参观校园，介绍学院主要建筑、介绍中德技术学院、指导他乘车出行等。

Dialogue 1　Meeting the New Student 遇见新同学

Z: Welcome to our college. I'm the **volunteer**. My name is Zhang Yang. Let me help you with your **luggage**.①	义工 行李
W: Thank you very much. It's very kind of you.	
Z: **It's nothing**.② May I have your name?	别客气
W: I'm Wang Nan. Nice to meet you.③	
Z: Nice to meet you too, Wang Nan. I'm from Sino-German Technical College. I major in④ **CNC Technology**.	数控技术
W: Wow! We are in the same department. My major is **Industrial Process Automation Technology**.	工业过程自动化技术
Z: Good. Since you are a **freshman**⑤, later I will show you around and you'll **be familiar with our campus**.	新生 熟悉我们的校园
W: That's so kind of you!	
Z: Where are you from?	
W: I'm from Qingdao. What about you?	
Z: Small world! I'm from Qingdao **as well**!	也
W: Great! We are from the same place, we should help each other in the future.	
Z: That's for sure. Here we are at the **dormitory**. Let me help you settle down. The school **canteen** is not far away. There's a **supermarket** to its right.	宿舍 餐厅，超市
W: Thank you so much for your kind help.	
Z: You're welcome!	

Note:

① 此处是同学之间的自我介绍和向别人提供帮助的说法。在比较正式的场合或向师长、上司自我介绍前，可以先客气的问一句"May I introduce myself？"，后一句中的"May I have your name？"是询问别人的姓名时的客气的说法，意思等同于"What's your name？"。向别人提供帮助可以说"Let me help you with..."或者询问"What can I do for you？"。

② "–Thank you very much. It's very kind of you. –It's nothing."是致谢和回复的常用说法。通常在表达感谢时，根据情境在 thank you 之外加上修饰成分，会使表达更具体。常用的感谢的表达还有"Thank you so much for your kind help.""I do appreciate all you've done for me.""Thank you for your attendance.""Thank you anyway."回答除了"It's nothing."还有"You are welcome.""That's all right.""My pleasure."。

③ 两位同学初次见面，用"Nice to meet you"互相问候。在正式场合，初次见面，两人还可以说"How do you do？"非正式场合，也非第一次见面时，在欧美学生间比较流行的问候语是"How are you doing？"，常用的回答包括"Great""Pretty good"等。

④ major 既可以作动词也可以作名词，都是主修、专业的意思，除了对话中的两种用法外，"我学某专业"还可以说"I'm a ... major."。

⑤ freshman 指大一新生，大学二年级学生可以叫做 sophomore，三年级学生是 junior，四年级学生则被称为 senior。

Activity 2: Oral Practice 口语练习。

Please extend greetings to one of your classmates, and introduce yourself.

Activity 3: Fill in the blanks according to the dialogue. 根据对话内容填空。

1. Zhang Yang is a _____ major, while Wang Nan majors in _____.
2. Both Zhang Yang and Wang Nan come from _____.
3. Zhang Yang introduces the position of _____, _____ and _____.

Dialogue 2　Visiting the Campus 参观校园

Z: Hey, Wang Nan. Let's go around our campus.	
W: Great! The campus is very big and beautiful.	
Z: That's right. Now we arrive at the **Teaching Building**.	教学楼
W: Don't you mean that many of our **courses** will be taken here?	课程
Z: Yes. The **theory-learning** will be taken here. And the theoretical knowledge learned will be **put into practice** in the **workshop**.	理论学习 付诸实践，车间
W: So the **practical training** will be taken in the workshop.	实训
Z: Right! The workshop is over there. Let's go and have a look.	
W: Yeah, then we will be more **competitive** in the job market.	有竞争力
Z: Definitely! Due to the "**Dual Education System**", the college and German companies will work hand in hand① to **train** us.	双元制教育 培训
W: So we are **apprentices** as well as② students.	学徒
Z: After **graduation**, some students get promoted③ and become **team leaders** or **trainers** after several years of **diligent** work.	毕业，班组长 培训师，勤奋的
W: Fabulous④! It seems that the workshop is so **professional**.	专业的
Z: You will be attracted by the machines when you operate them.	
W: I can't wait to⑤ work with them.	

Note:

① hand in hand 是手拉手，联合的意思。例如，The two departments work hand in hand to solve the tough problem. 两个部门联手解决这个棘手的问题。

② as well as 的含义是"还有""不但……而且……"。值得注意的是，在 A as well as B 的结构里，语意的重点在 A，不在 B。因此，"We are apprentices as well as students." 的译文应该是："我们不但是学生，还是学徒。"

③ get promoted 的意思是"得到提升"，promotion 是其名词形式。

④ fabulous 的意思是"极好的，非常精彩的"，例如：look fabulous 表示极好看，feel fabulous 表示感觉好极了。

⑤ can't wait to do sth. 表示迫不及待要做某事，例如：The boy can't wait to play computer games. 那个男孩迫不及待要玩电脑游戏。

Activity 4: Please answer the following questions according to the dialogue. 请根据对话回答问题。

1. What are the 2 learning places at college mentioned in the dialogue?
   _____

2. What are the 2 kinds of knowledge that will be learned mentioned in the dialogue?
   _____

3. What are the 2 parties of teaching mentioned in the dialogue?
   _____

4. What are the 2 roles of students mentioned in the dialogue?
   _____

5. Could you try to summarize the meaning of "dual" in the dialogue?
   _____

Activity 5: Look up dictionaries and figure out the meanings of the following phrases. 查字典写出以下短语的含义。

1. face to face _____
2. back to back _____
3. shoulder to shoulder _____
4. eyes to eyes _____
5. heart to heart _____
6. side by side _____
7. arm in arm _____

Dialogue 3 Asking the Way 问路

Z: This is the canteen. The **playground** is next to the canteen. Many students play football here after class.	操场
W: Great! That's **my favorite sport**!	我最爱的运动
Z: I'm a **member of the college football team. Do you want to join us**?	校足球队队员
W: Sure. How can I join it?	= You wanna join us?
Z: You can apply to① the **student union**.	学生会
W: That's great! I'll go and apply for① the **membership** tomorrow. By the way, could you tell me how I can get there②?	会员身份
Z: It's on the 3rd floor of the library. You need to go along this way, turn left at the second **crossing** and you can see the Panlong **Square**, the library is on your left. Go upstairs to the third floor and the student union is the second office on your left.	路口，广场
W: Then would you mind③ telling me the way to the city center?	
Z: You'd better catch a bus. You can take Bus 306, 102, or 112 at the school gate. And the **bus stop** is on the other side of the road. It's about ten minutes' walk.	公交车站
W: How can I get to the **subway**④ **station**?	地铁站

Z: All the three buses go to the station. Just remember not to **miss the direction**.	弄错方向
W: Fine. Now I have got a **general picture** of our school: library, canteen, playground, bus stop... Thanks a lot!	大致印象
Z: My pleasure. We'll spend three nice years on this campus.	
W: It's 5 o'clock. Let's go for dinner together for your help and kindness. **It's my treat.**	我请客
Z: Many thanks. Let's go.	

Note:

① apply to 的宾语是申请的对象，例如经理 (manager, director)，大学 (university, college) 等，而 apply for 的宾语是希望获得的事物，例如工作 (job, work)，职务 (post, position)，奖金 (reward)，签证 (visa)，工资的提高 (an increase in salary)。

例如，He has applied to the banker for loan. 他已向这位银行家申请了贷款。For further information, apply to the secretary of the company. 欲获得进一步的详情，请与公司秘书联系。

I am applying for a job as a clerk. 我正在申请一个做职员的工作。Nobody applied for the reward. 谁也没有申请这项奖金。

② 问路常用的句型如下：Could you show me the way to the workshop?/ Would you mind telling me the way to the workshop? /How can I get to the workshop?

③ would you mind doing 这个句型用来表示请求，意思是"请你做……你是否介意？""请你做……好吗？"是一种比较客气的表达方式。例如：Would you mind turning off the light in the room? 请你把房间里的灯关掉好吗？

要表示"请你不要做……你是否介意？""请你不要做……好吗？"只需要在 doing 前面加上 not。如：Would you mind not standing in front of me? 请你不要站在我的前面好吗？

回答时，如果同意对方的请求，可用"Certainly. / Of course not. / Not at all. / No, not at all."来表达；

如果不同意对方的请求，常用"Sorry. / I'm sorry."及陈述某种理由来表示拒绝或反对。

例如：

—Would you mind going to the movies this evening? 今晚去看电影好吗？

—I'm sorry. But I haven't finished my homework yet. 对不起，我的作业还没有完成。

Would you mind doing...?"句型中的逻辑主语只能是谈话的对方 you。如果想要对方允许自己做某事，可用"Would you mind my doing...?"句型，如：Would you mind my smoking here? 你介意我在这里吸烟吗？

④ subway 地铁，其中 sub 是英语中常用的前缀。意思是"在……之下的"，例如，subway（地铁），submarine（潜水艇），subconscious（潜意识）。有时 sub 也表示级别上低一等的，例如，subtitle（副标题），subassembly（次组合件）。

前缀由一个字母或一组字母组成，附在单词的开头，用于表明或修饰词义。理解不同前缀的含义是很重要的，因为这有助于理解所学的新词汇的含义。要注意的是，前缀有时有不止一个意思！另外，如果将前缀去掉，即使词义发生变化，该词仍然是一个完整的词汇。

常见的前缀有：a. 表示正负（或增减）的，如：un-,in-,im-,non-,mis-,dis-,anti-,re- 等；b. 表示尺寸的，如：semi-,mini-,micro-,macro- 等；c. 表示位置关系，如：inter-,super-,trans-, sub- 等；d. 表示时间和次序，如：pre-,post- 等；e. 表示数字的，如：semi-, bi-, tri-, quad-, multi- 等。

Activity 6: Word Game. Reorder the letters to make correct words of places at school and write down their meanings in Chinese. 词汇游戏。重新排列字母组成与校园有关的单词并写出它们的汉语意思。

LETTERS	WORDS	MEANINGS
1. arndoulgpy	p_____	_____
2. etennca	c_____	_____
3. timodroyr	d_____	_____
4. phsrkowo	w_____	_____
5. lrrayib	l_____	_____
6. qersau	s_____	_____
7. mpsuac	c_____	_____
8. partemusrek	s_____	_____

Activity 7: Complete the following sentences using the phrases. 用提供的短语完成句子。

1. Would you mind _____? (help me wash my clothes)
2. Would you mind _____? (give her a cup of tea)
3. Would you mind _____ my baby while I'm away? (look after)
4. Would you mind _____ her a message? (give)
5. Would you mind my _____ the window? (open)
6. Would you mind _____ me some suggestions on Math learning? (give)

Activity 8: Change the following words by adding proper prefixes. 给下列单词加上适当的前缀。

1. media（媒体）_____（多媒体）
2. necessary（有必要的）_____（没必要的）
3. agree（同意）_____（不同意）
4. large（大的）_____（扩大）
5. lead（带领）_____（误导）
6. clockwise（顺时针）_____（逆时针）
7. active power（有效功率）_____（无效功率）
8. corrosion（腐蚀）_____（防腐蚀）
9. assembly（组件）_____（次组合件）

10. potential（电位）_____（等电位的）

❋ Reading and Writing 读写拓展

Reading 阅读

FESTO (FESTO Production Ltd.), the global leader in bionic technology and pneumatic automation field, is of decisive importance in Chinese pneumatic industry and wins the praise of the industry with its remarkable quality, excellent problem-solving skills and perfect service.

The headquarter of FESTO is located in Esslingen, which is the capital and largest city of Esslingen, the Stuttgart district of Baden-Württemberg, Germany. It is located on the banks of the Neckar River, 14 km southeast of Stuttgart, with an area of 46.43 square kilometers and a population of 92,299. Neckar River cross the city from the southeast to the northwest and the old city is located in the north bank of the river.

bionic [baɪˈɑnɪk] adj. 仿生学的；利用仿生学的
pneumatic [numætɪk] adj. 气动的；充气的；有气胎的
remarkable [rɪˈmɑːkəb(ə)l] adj. 非凡的；卓越的
excellent [ˈeksələnt] adj. 极好的；杰出的
perfect [ˈpɜːfɪkt] adj. 完美的
quality [ˈkwɒlətɪ] n. 质量
problem-solving skills 问题解决能力
service [ˈsɜːvɪs] n. 服务
headquarter [ˈhedkwɔːtə] n. 总部

Activity 9: Read the following passage and finish the questions below. 读短文回答问题。

It was my first day on campus. The Winona State University of Minnesota, a small but neat and clean school. I flew a long way from Bangladesh to be here. The whole place looks white from the plane as it was covered with snow. I knew from my brother that it's very cold in Minnesota as it is on the north. So I covered myself with warm clothes.

I was nervous to be in a new place, but kind of curious too. A Bangladeshi student of WSU came to pick me up. I found myself in a new world, new people.

It's impossible now to express those feelings in words here. But I can tell you my experiences. I felt little comfortable when I saw some Bangladeshi students in the orientation. They were the representative of Bangladeshi student organization. I met students from different countries, like China, Japan, Africa, Iraq. I met one guy from Palestine. Time were moving with joy, I met a beautiful girl from South America who was really a nice person.

We got a lunch break. First time in my life I got the opportunity to taste American food which looks nice but didn't fulfill my appetite as I love rice and curry. But I know I have to depend on these foods. After lunch we were taken to visit the campus. Our group leader took us to different buildings and explained clearly. By this time I made some friends who are from Malaysia. I was enjoying every minute. We were also taken to a building to take pictures for our school ID. I asked my group leader who was a Bangladeshi to take me to my dorm. I was kind of scared first as my roommate is an American. But when I met Riley (My roommate) I felt like he is one of us.

The sun went to bed soon and I felt tired as I walked a long way in the snow and it's freezing outside. I prepared my mind for the next day as I have to go to my first class at 8:30 in the morning. So time to go to bed.

Such is my first day on campus.

1. How did he get to Winona State University of Minnesota?
   A. By train.　　　　B. By bus.　　　　C. By plane.　　　　D. By sea.
2. What does the word "appetite" (Para.4, Line 2) mean?
   A. Desire to eat　　　　　　　　　　B. Strong desire for success
   C. Goal　　　　　　　　　　　　　　D. Feeling
3. Which country is the writer from?
   A. Palestine.　　　B. Bangladesh.　　　C. America.　　　D. Malaysia.
4. Which of the following activity is not mentioned in the passage?
   A. Take ID card photo.　　　　　　B. Arrive at the dorm.
   C. Taste rice and curry.　　　　　　D. Visit the campus.
5. Which is the best title for the passage?
   A. I Entered a New College　　　　B. Various People of My College
   C. How Did I Feel in the New University　　D. My First Day on Campus

Writing 写作

Leave/Absence Request Form 请假申请

日期：
Date:

姓名 Name		部门 Department		职务 Title	
1. 请假类别：　□事假　Personal Leave　　　□病假　Sick Leave 　　　　　　　　□婚假　Marriage Leave　　□年假　Annual Leave 　　　　　　　　□工伤　Work-related Injuries　□产假　Maternity Leave 　　　　　　　　□护理假　Nursing Leave　　□其他　Others					
2. 请假事由： 　　Reason:					
3. 请假时间：自_____至_____，共计___天 　　Date:　　From_____To_____, ___ days in total					
总经理 General Manager			部门经理 Dept. Manager		

Request for Leave 请假条

The Request for Leave is one of the commonly used applications. The commonly used ones are

Request for Sick Leave and Leave of Absence.

There are three points you need to pay attention when writing a request for leave:

1. The request for leave has many similarities with letters, which is a great simplification of letters. It is generally composed of four parts: time, title, text and signature.

2. The first paragraph should begin with polite request for leave. The second paragraph should explain the reason for the leave and apologize. At the end, express your wish that you want to get a leave.

3. The feature of it includes straightforward, brief, and easy to understand.

请假条是常用的申请之一。最常见的包括病假和事假。

在写请假条时需要注意三点：

1. 请假条与信件有很多相似之处，是对信件的简化。它通常由四部分组成：时间、标题、正文和签名。

2. 第一段应以礼貌的请假开始。第二段应说明请假的原因和道歉。最后，再次表达你想要请假的愿望。

3. 请假条的显著特点是直接、简单明了、易于理解。

Example:

To: John Smith, Supervisor
From: George Chen, Accounting Department
Date: May 9th, 2018
Subject: Personal Leave

Dear Mr. Smith,

I would like to know if I could ask for a personal leave for one day on May 18th.

This morning I received a letter from a friend, inviting me to attend his wedding ceremony which will take place next Friday, May 18th. I am so pleased to hear this good news that I cannot wait to see my best friend become the happiest groom in the world.

I hope a one-day leave next Friday will not cause much trouble to you. You will be highly appreciated if you grant me my request.

Yours,
Wang Nan

Note：

常见的英文书信格式有两种。一种是传统的缩排式（indented style），即每个段落的第一行要空格，结束语、签名等也要缩排，例如 Activity 10 中的请假条是典型的缩排式。上文中的格式是更简洁的齐头式（block style）。在齐头式中，信中的每个部分都从左边界起头，不同的部分之间用空行分隔开。由于这种格式在写电子邮件时更方便，所以更受欢迎。

Activity 10: Please write a Request for Leave according to the following context. 请根据下面的情境写一份请假条。

Context：王楠由于昨天 (2018 年 4 月 5 日) 淋了雨发烧了，所以今天第一二节课需要请假，请姜老师批准。

```
To: _____
From: _____
Date: _____
Subject: _____

Dear_____,
    I _____（非常抱歉）that I am unable to come to the 1st period today.
    Due to_____（昨天我淋了雨，晚上发烧了）. The doctor advised me to rest for several days and after my recovery, I will attend school as soon as possible.
    I am _____（非常抱歉）about the inconvenience. However, I would like to have your_____（批准）. Also, afterwards I would phone you and there's the certificate by the doctor.

Yours,
Wang Nan
```

Speaking Skills 交际技巧

妙词巧句

话到嘴边说不出来？牢记一些日常对话中的句式是你生活中的一把必备钥匙，有时一个简单的句子就能让我们的英语听起来更地道。平时注意积累一些简短而含义丰富的短句或习语，对提高英语口语交际水平很有帮助。

1. You're in the pink! 形容人气色好，除了"You look fine!"，加上些表示颜色的词能使得句子更形象生动。

2. He is bouncy. 他精力充沛。
美国人通常不用 He is energetic.。

3. I get mind of you. 久仰。
I heard a lot about you. (×)

4. I would rather not say. 还是别说了吧。
当别人问你不愿公开的问题，切勿用"It's my secret, don't ask such a personal question."。

5. It's on the tip of my tongue.
有时候，你想说什么但想不起来，可以说"Well""Let me see.""Just a moment."或"It's on the tip of my tongue."。相比之下，最后一个句型也是最地道的。

6. While I remember...
交谈时，如需转换话题，不要只会说"By the way"，实际上，"To change the subject""Before I forget""While I remember...""Mind you"都是既地道又受欢迎的表达。

7. I got it.
I know 可能是被中国人用得最多，而又最不被美国人接受的一句话。当一个美国人向你解释某个问题时，你如果连说两遍"I know"，他就不会再跟你说什么了。这时，"I got it"就顺耳得多。如果仍然有问题不明白，可以用"I'm not clear about it." "It's past my understanding." "It's beyond me."。

8. Go down to business. /Get to the point. 言归正传。
9. Give him the works. 给他点教训。
10. I'll see to it. 我会留意的。
11. You have my word. 我保证。这是除了"I promise"的另一种地道的用法。
12. Great minds think alike. 英雄所见略同。
13. First come, first served. 先到者先得。

Activity 11: Please match the expressions in these two columns by the same meaning. 请将下列两列句子中相同意思的句子连线。

1. She looks blue today.   A. Focus your attention!
2. Be there be square.   B. I'm not leaving until I see you.
3. In your dreams.   C. It's impossible to take place.
4. None of your business!   D. She seems depressed today.
5. Stay on the ball.   E. Stay out of my business!

Self-evaluation

Items	Yes	No
I have acquired all the key words in the Mission Objective		
I have acquired all the sentence patterns in the Mission Objective		
I can compose a Request for Leave		

Unit 2
Major Introduction 专业介绍

 Warming-up 课程准备

Mission Objective 任务目标

1. 掌握机械工程师的主要职责、工业过程自动化技术专业和数控技术专业主要学习领域、培训基地、理论学习、实训、高素质技术工人、安装、气体、液体、装配、调试、维修、劳动力、工作流程、蓝领、金蓝领等的英文表达。
2. 会用英语说明机械工程行业的重要性、描述双元制教育的特色和优势、表示称赞。
3. 会撰写英文会议记录。

Discussion 讨论

1. Can you say your major in English?
2. Can you guess the top 10 university majors?
3. Can you list your former classmates' majors?

Words and phrases for assistance:

专业、技能、职业：Industrial Process Automation Technology, CNC Technology, make parts, operate machines, design parts, inspection, a fitter, CNC operator, trainer, team leader, practical skills, theoretical knowledge

Database 知识库

Mechanical Engineering

Mechanical Engineering is closely related to the development of the national economy. It provides mechanical equipment and electrical equipment for all walks of life, known as the "Equipment Department of the national economy". It not only strengthens the application of basic knowledge on Mathematics and Mechanics, but also bases on the design, processing and manufacturing of mechanical structures, and integrates automatic control technology, information technology, computer technology, art and so on. It studies and solves the theoretical and practical problems in the development, design, manufacture, installation, application and maintenance of various machinery. At the same time, the development of many industries is inseparable from the technical support of Mechanical Engineering, such as aerospace, construction machinery, agricultural machinery and so on.

Mechanical Engineering includes majors such as CNC Technology, Industrial Process Automation Technology, Industrial Design, Mechatronics, Mechanical Design and Manufacture, Automobile Maintenance, etc.

Employment: Mechanical Engineering is a major discipline in Engineering. As long as we use equipment and production lines, we need mechanical engineering knowledge. In addition to teaching and marketing, positions such as production directors, logistics management, equipment management, quality management, project management, mechanical product development, automobile industry, mold design and manufacturing, CNC engineers are available to Mechanical Engineering majors. They can also engage in highly skilled posts (such as equipment maintenance, CNC maintenance, environmental protection equipment design, etc.).

"Mechatronics" is a term coined by the Japanese to describe the integration of mechanical and electronic engineering. The concept may seem to be anything but new, since we can look around us and see a myriad of products that utilize both mechanical and electronic disciplines. Mechatronics, however, specially refers to a multidiscipline, integrated approach to product and manufacturing system design. It represents the next generation of machines, robots, and smart mechanisms necessary for carrying out work in a variety of environments—primarily, factory automation, office automation, and home automation.

Activity 1: Fill in the blanks according to the passage. 根据短文内容填空。

1. Majors of Mechanical Engineering includes _____, _____, _____, _____, _____, _____ etc.
2. Mechanical Engineering studies the _____, _____, _____, _____, _____ and _____ of various machinery.
3. The term Mechatronics comes from _____.
4. Mechatronics seems to be a concept which is not _____. However, it specially refers to a _____, _____ approach to product and system design.

Activity 2: Answering Questions. 回答问题。

Can you summarize the importance of Mechanical Engineering?

Oral Practice 口语训练

人物涉及 (Characters):

D: Mr. Dittrich (German expert of Sino-German Technical College) 中德技术学院德方专家

L: Mr. Li (Training Manager of FESTO) 费斯托公司培训经理

C: Mr.Chen (Director of Sino-German Technical College) 中德技术学院主任

W: Wang Nan (Freshman Majoring in Industrial Process Automation Technology) 工业过程自动化技术专业新生

G: Gao Jian (Freshman Majoring in CNC Technology)　数控技术专业新生
Z: Zhang Yang (Signed Apprentices of FESTO)　费斯托签约学徒

情景解说 (Situations)：

费斯托公司将在济南投资建厂，培训经理 Mr. Li(L) 电话联系中德技术学院德国专家 Mr. Dittrich(D) 了解双元制学徒的培养情况并准备先选拔一批学员。

新生 Wang Nan(W) 向中德技术学院主任 Mr. Chen(C) 咨询大学三年的学习内容。

新生 Gao Jian(G) 与签约学徒 Zhang Yang(Z) 交流。

以下对话围绕这三个情境展开。

Dialogue 1　Consultation through Telephone 电话咨询

(Mr. Dittrich's phone is ringing)

D: Hello! This is Dittrich!	
L: Hi, it's Li Ning speaking, training manager of FESTO. **I made the appointment** with you last week for today's communication①.	预约
D: I am glad to talk with you. What can I do for you?	
L: I'd like to have some information of the **training base**.	培训基地
D: Sure. The **Delegation of German Industry & Commerce** in China–Shanghai (AHK Shanghai), the Jinan Vocational College and several German companies **established** the training center in 2011. Now it is the seventh year after its **foundation**. The aim is to **educate high–quality** skillful **personnel** with "Craftsman Spirit" based on German standards②.	德国商会 创办 教育，高素质的，职员
L: How many majors do you have **at present**?	目前
D: We've got two majors. 50 students major in **CNC Technology** and 50 students major in **Industrial Process Automation Technology** per year③.	数控技术专业 工业过程自动化技术
L: What about the **study period**?	学习年限
D: 3 years. The students' ages are between 18–22.	
L: What are the students' **advantages**④?	优势
D: The **mode of Dual Education** is a model of success. The apprentices have already done practical courses in the companies, so they know best about **company requirements**. The companies save time and money because the **skilled workers** are already **integrated**⑤. Are you interested?	双元制教育模式 公司要求 熟练工 整合的
L: It sounds great⑥. How many **cooperated companies** do you have now?	合作企业
D: We have more than 10 cooperated companies. Over the past ten years, many German production companies have **set up facilities** in the east of China. A major **bottleneck** for German companies' growth in China is a **shortage** of **qualified labor force, blue–collar workers**, especially technicians of Industrial Process Automation Technology.⑦	建立，设备厂址 瓶颈，短缺 合格的劳动力，蓝领工人
L: I'd like to select 4 students from the **sophomores** as our apprentices.	大二学生
D: OK, you can e-mail the **position requirement** and your company's introduction to me.	职位要求

L: I will **arrange** it as soon as possible.	安排
D: OK, we can make sure the **interview** time later.	面试
L: Before interview, I will go to the college for **face-to-face** communication, and **make the schedule**.	面对面 确定时间表
D: Welcome to visit the college.	
L: Thank you for your information. Bye.	
D: You are welcome. Bye.	

Note:

① Hi, it's Li Ning speaking, training manager of FESTO. I made the appointment with you last week, for today's communication. 您好！我是费斯托公司的李宁。上周跟您预约过今天的交流。

② Sure. The Delegation of German Industry & Commerce in China-Shanghai (AHK Shanghai), the Jinan Vocational College and many German companies started the training program in 2011. Now it is the seventh year after its foundation. The aim is to educate high-quality skillful personnel with "craftsman spirit" based on German standards. 中德培训项目始于2011年，是由德国工商大会上海代表处、济南职业学院以及德国企业共同成立的，已经运行了七年。主要是为德国在华企业培养符合德国标准的具有工匠精神的高素质技能型人才。

③ per 在文中是"每一……，每个……"的意思。例如，per day, per week, per person, per meter 等。as per 是在说明书等材料中常见的短语，是"按照……，根据……"的意思，例如 as per instructions, 按照说明。

④ 关于优缺点的几种表达形式：

优势：advantage, merit, virtue, benefit, upside, strength

应用句型：People have generally discovered several advantages in..., of which the most significant/obvious one is.../that...……有许多优点，其中最有意义、最明显的一个是……/从句。

缺陷：disadvantage, demerit, drawback, downside, weakness

应用句型：In the meantime, we can not ignore the disadvantages of... 同时，我们不能忽视……的不利方面。

⑤ The mode of dual education is a model of success. The apprentices have already done practical courses in the companies, so they know best about company requirements. The companies save time and money because the skilled workers are already integrated. 双元制教育模式是成功的。学徒在公司进行实训，他们非常了解公司要求。公司节约时间和成本，因为这些都是熟练工。

⑥ It sounds great! 的意思是"听起来不错；听起来很棒"。Sound 在这里是系动词，它本身有词义，但不能单独用作谓语，不能用进行时态，后边必须跟表语（adj）即：link v + adj，构成系表结构说明主语的特征、性质、状况等。类似的系动词还有 feel, smell, look, taste 等。例如，This kind of cloth feels very soft. 这种布手感很软。This flower smells very sweet. 这朵花闻起来很香。

⑦ We have more than 10 cooperated companies. Over the past ten years, many German production companies have set up facilities in the east of China. A major bottleneck for German companies' growth in China is a shortage of qualified labor force, especially technicians of Industrial Process Automation Technology. 现在我们有超过10家合作企业。过去十年，许多德国制造公司在华东地区建厂，目前这些企业在中国发展的主要瓶颈是合格劳动力的短缺，特别是工业过程自动化技术方面的技师。

set up：创建，建立，组建。eg. They set up many branches throughout the country. 他们在全国各地组建了很多分公司。

Activity 3: Filling in the blanks using the phrases in the table. 用下列短语完成句子填空。

set up	set down	set back	set about

1. Passengers may be _____ and picked up only at the official stops.
2. The bad weather _____ the building program by several weeks.
3. We need to _____ finding a solution.
4. He _____ a stand on the pedlars' market.

Activity 4: Choose the best answer from the four choices. 选择恰当的答案。

1. –Have you heard of the story? It _____ funny and interesting.
   A. looks   B. tastes   C. smells   D. sounds
2. –Dinner is ready. Help yourself!
   –Wow! It _____ delicious. Could you tell me how to cook it?
   A. looks   B. tastes   C. smells   D. sounds
3. –Do you like the material?
   –Yes, it _____ very soft.
   A. is feeling   B. felt   C. feels   D. is felt
4. She looks _____ .
   A. happy
   C. happily
   B. to be happy
   D. that she is happy

Activity 5: Oral practice 口语训练

1. Which do you think is more important, practice in the workshop or classroom learning? Please give your reasons.
2. What are the challenges for the international companies in China?

Dialogue 2   Industrial Process Automation Technology 工业过程自动化技术专业

W: Good morning, Mr. Chen. What can I learn during the 3-year study period? C: Good morning. Our **curriculum** is **distinguished**. The training plan is **classified** into three parts, which are **Professional Quality**, **Professional Development**, and **Professional Skill**[①]. W: Can you give me more information of Professional Skill?	课程，有特色的，分为 职业素养，职业发展 职业技能

C: Yes, it mainly **focuses on thirteen learning field**, the same as German Industrial Process Automation Technology② **technician** training. We have courses like: Electric **Installation**, Gas-liquid Transmission Control Technology, Data Processing, Machine **Assembling**, **Adjustment** and **Maintenance**, etc.③	集中于 13 个学习领域 技师 安装，气体—液体 装配，调试 维修
W: Sounds good! Is it hard to learn?	
C: **Where there is a will there is a way**. Teachers will organize **working tasks** for you. We encourage you to learn by work, and to work in order to learn④.	有志者事竟成，工作任务
W: Interesting⑤. I will be **bored** if I always stay in the classroom.	厌烦
C: Ah, it seems you like your major and you like doing things by hands. The theory is integrated with the working tasks. You will not only stay in the classroom to learn theory.	
W: It's great. Thank you.	
C: You are welcome.	

Note:

① Our curriculum is distinguished. Every major's training plan is classified to three parts, which are Professional Quality, Professional Development, and Professional Skill. 我们的课程是很有特色的。每个专业的培养计划都分为职业素养、职业发展和职业技能三个部分。

② Industrial Process Automation Technology 工业过程自动化技术。It is a multidisciplinary field of science that includes a combination of mechanical engineering, electronics, computer engineering, telecommunications engineering, systems engineering and control engineering. 这是一门多学科的科学领域，它包括机械工程、电子学、计算机工程、电信工程、系统工程和控制工程。

③ We have courses like: Electric Installation, Gas-liquid Transmission Control Technology, Data Processing, Machine Assembling, Adjustment and Maintenance, etc. 专业课程包括电气安装、气液传输控制技术、数据处理、设备安装、调试与维护等。

④ Where there is a will there is a way. Teachers will organize the working tasks for you. We encourage you to learn by work, and to work in order to learn. 有志者事竟成。每个学习领域老师都会组织很多个工作任务，在工作中学习，学习中工作。

⑤ 以后缀 –ing 结尾的形容词 (如 delighting, exciting, frightening, interesting, surprising, worrying 等) 主要用于说明事物，表示事物的性质或特征。-ed 结尾的形容词 (如 delighted, excited, frightened, interested, surprised, worried 等) 通常直接用于说明人。例如，I was bored in the boring maths lesson. 在枯燥的数学课上我感到无聊。

Activity 6: Match the 13 learning fields with their Chinese equivalents. 将下列 13 个学习领域的英语名称与汉语名称连线。

1. Analyze Functional Correlations in Mechatronics Systems
2. Produce Mechanical Sub-systems
3. Install Electrical Equipment according Due Consideration to Technical Safety Aspects
4. the Energy and Information Flows in Electrical, Pneumatic and Hydraulic Sub-assemblies
5. Communicate with the Assistance of Data Processing Systems
6. Plan and Organize Work Processes
7. Realize Simple Mechatronics Components
8. Design and Develop Mechatronics Systems
9. Investigate the Information Flow in Complex Mechatronics Systems
10. Plan Assembly and Disassembly
11. Commissioning, Trouble shooting and Repair
12. Preventative Maintenance
13. Hand over Mechatronics Systems to Customers

A. 机械子系统的制造
B. 安全保障下的电气设备安装
C. 电能与信息流的查验、气动与液动组件
D. 使用计算机进行通信
E. 机电系统的功能关系分析
F. 机电一体化系统的移交（客户）
G. 装配与拆卸计划
H. 复杂机电一体化系统（中）信息流（信号）的调查（测试）
I. 机电一体化系统的设计与生产（建立）
J. 调试、故障诊断和维修
K. 机电一体化子系统的实现
L. 预防性维护
M. 工作过程的计划与组织

Activity 7: Choose the best answer from the four choices. 选择恰当的答案。

1. As we all know, typing is a _____ job.
   A. tired　　　B. tiring　　　C. tire　　　D. to tired
2. Laws that punish parents for their little children's actions against the laws get parents _____.
   A. worried　　　B. to worried　　　C. worrying　　　D. worry
3. He was _____ to see Helen.
   A. surprised　　　B. surprise　　　C. surprising　　　D. to surprise
4. If you call something _____, you mean it _____ you.
   A. interesting, interests
   B. interesting, interesting
   C. interests, interesting
   D. interested, interested

Dialogue 3　CNC Technology 数控技术专业

Z: Gao Jian, why did you choose **CNC Technology** as your major? G: Many factors led me to major in CNC Technology. The most important factor is that I like **tinkering with** machines. Z: In fact, CNC Technology is far more complex than tinkering with machines[①]. The courses in the major include Machine and **Programming**, Automatic Control, CNC Operation, Mechanical Drawing, PLC, etc[②].	数控技术 修补 编程

G: Oh, that's attractive! I hope **practical training** is as much as **theory-learning**.	实训，理论学习
Z: Ah, Dual Education System provides exactly ③ what you want. It pays much attention to practical training and we'll spend much time in workshop.	
G: What can we do as ④ an apprentice in the company?	
Z: During the **internship**, the apprentices should be **front-line workers** ⑤ first. We'll experience different duties in different departments like **Production Department, Logistics Department, Maintenance Department** and so on. I am a **fitter** in the Production Department now.	实习，一线工人 生产部 物流部，维修部 装配工
G: Then we'll get familiar with all the **working procedures** in the departments.	工作流程
Z: You are right. ⑥ That's the shortcut to become a **golden blue collar**.	金蓝领

Note:

① far more complex 中, far 用来修饰比较级 more, 表示"远远的多于……"例如: This book is far more interesting than that one. 这本书要比那本书有趣的多。

② The courses in the major include Machine and Programming, Automatic Control, CNC Operation, Mechanical Drawing, PLC, etc. 专业课程包括机床与编程、自动控制、数控机床操作、机械制图、PLC 等。

③ exactly 是副词，用在这里表示强烈的语气，意思是完全符合事实或标准。句子中使用各种副词加强语气，也是使口语表达更地道的好方法。

④ as 在这里用作介词，作"充当，作为"的意思。例如，As a writer, he is famous. 作为作家，他是很有名的。

⑤ 机械专业一线工人也叫"first-line worker"，可以细分为：machine man(机械工)，locomotive man (机车工)，metaler/locksmith(钳工)，miller/milling machine operator(铣工)，grinder(磨工)，driller/drill man(钻工)，lathe man(车工)，carpentry(木工)，electrician(电工)。

⑥ 本单元的三篇对话中有多处使用了表达称赞的常用语。包括"It sounds great!""Sounds good!""Interesting!""It's great!""That's attractive!"。

Activity 8: Complete the sentences according to the Chinese. 根据所给汉语完成句子。

1. Health is _____ than wealth and wisdom. （重要的多）

2. Flying kites in the open air is _____ than playing chess. （有趣的多）

3. Bosses are _____ to hire someone who is like himself. （更有可能）

4. The Mayan calendar is _____ than the European calendar. （更准确）

5. It is _____ to travel by plane than by train. （贵得多）

Activity 9: Oral practice. 口语训练

1. What is your ideal job in the future?

2. Do you think to be a front-line worker is boring? Why or why not?

Assignments: 课后任务

Do you think the university students should have a part-time job?

Reading and Writing 读写拓展

Reading 阅读

Harsewinkel is a town in Gütersloh District in the state of North Rhine-Westphalia, Germany. It lies on the river Ems, some 15 km north-west of Gütersloh. It is the home and domicile of Europe's leading combine harvester manufacturer CLAAS, which is a major employer in the town.

CLAAS (CLAAS Agricultural Machinery (Shandong) Co. Ltd.), the German company of over 100 years of history, is mainly engaged in the production of the combines, forage harvesters, tractors and other agricultural equipment.

manufacturer [ˌmænjuˈfæktʃrə] n. 制造商，厂商
engaged in 从事，忙于
combine [kəmˈbain] n. 联合收割机
forage harvester 牧草收割机
tractor [ˈtræktə] n. 拖拉机
agricultural equipment 农业设备

Activity 10: Read the following passage and finish the questions below. 读短文回答问题。

Kasparov: Chess Computers Beatable... For Now

Humans will continue to beat computers for years, but the machines are likely to dominate in matches played over several games, according to the world's top chess player.

"We will not see a machine replacing a human being in our lifetime. Man will be able to beat a computer in at least one game for a very long time," Kasparov told a press conference in Moscow a week after settling for a draw in a six-game match with the computer Deep Junior in New York. But while human intuition can provide an advantage in individual games, "Man will never be able to play 8 or 10 games in a row to an equal level," Kasparov said. He gradually declines for a variety or external factors: the weather, a headache, family strains or whatever.

In his Man vs. Machine contest in the United States, Kasparov won the first game, but lost the third after committing a mid-game blunder. He then adopted a safety-first strategy, and in the sixth game passed up a chance to win by accepting a draw in a position some analysts said was favorable.

Kasparov-watchers believe he was determined above all not to lose to Deep Junior because he was still smarting from a defeat to another computer, Deep Blue, in 1997. That loss clearly rankled Kasparov, and he said at the time that the computer had been receiving assistance from its human operators.

The Russian, who has reigned undisputed as the world's top player since 1985, said he was "satisfied overall" with his result against Deep Junior, although, "if I'd been in better shape and had more time to prepare the result might have been different." He stressed the importance of psychology in chess between one human player and another, and described the "psychological discomfort" involved in adapting to a confrontation with a machine. "In chess with humans, you're always attempting to impose your decisions on the personality of your rival. A game is always an exchange of errors, of imprecision. It's psychology. There's never complete exactitude or purity in a game of chess," he said. "But playing against a machine, beyond a certain point, to win or even to save the game you have to play with absolute exactitude, which is not human quality. Knowing this specificity of your rival creates a psychological discomfort which is very difficult to overcome."

Kasparov was at pains to stress that his 1997 defeat was an aberration: "The main thing was to show that what happened then has nothing to do with the fight between man and machine. Any impartial specialist can see that Deep Junior is much stronger that Deep Blue. The real battle begins now."

1. According to Kasparov, _____

A. humans can beat computers in individual games.

B. computers will never take the place of human beings in games.

C. human beings can never beat computers in individual or series games.

D. human intuition plays an important role in games.

2. In the contest with Deep Junior in the United States, Kasparov _____

A. lost the game. B. won the game.

C. settled for a draw. D. left the game unfinished.

3. Which of the following statements is true about Kasparov's contest with Deep Blue in 1997?

A. He made up his mind to win Deep Blue.

B. He was smart enough to have beaten Deep Blue.

C. Deep Blue received human assistance.

D. Kasparov was unwilling to admit his defeat by Deep Blue.

4. According to Kasparov, a human vs. machine chess game may involve all the following qualities EXCEPT that _____

A. it involves psychological discomfort in the mind of the human player.

B. it demands the human player of absolute exactitude.

C. it creates an exchange of errors between man and machine.

D. it is difficult to overcome psychological discomfort.

5. Kasparov's remarks on his 1997 defeat imply that _____

A. man was no match to computer in intelligence.

B. Deep Blue was unbeatable.

C. Deep Blue also made blunders.

D. if he had made no blunders, he should have beaten Deep Blue.

Writing 写作

Conference Registration Form 会议签到表

会议时间 Time of Meeting:		
会议地点 Place of Meeting:		
会议主题 Subject of Meeting:		
部门 Department	职务 Title	人员 Participants
会议记录 Minutes		

Minutes 会议记录

Minutes are the official written record of the meetings of an organization or group. They typically describe the events of the meeting and may include a list of attendees, a statement of the issues considered by the participants, and related responses or decisions for the issues. The organization may have its own rules regarding the content of the minutes. The minutes of certain groups, such as a corporate board of directors, must be kept on file and are important legal documents.

会议记录是公司会议的正式书面记录。它们通常描述会议的事件，还包括与会者名单、与会者的问题陈述，以及相关问题的回应或决定。各公司会议记录内容的规则可能有区别。某些团体（如公司董事会）的会议记录是必须存档的，并且是重要的法律文件。

Minutes may be created during the meeting by a typist, who then prepares the minutes and issues them to the participants afterwards. Alternatively, the meeting can be audio recorded or video recorded, a group's appointed secretary may take notes, with minutes prepared later.

会议记录可能由打字员在会议期间创建，会后完成，然后发给与会者。会议期间也可能录音或录制视频，由秘书记笔记，会后完成会议记录。

Generally, minutes begin with the name of the body holding the meeting and may also include the place, date, list of people present, and the time that the chair called the meeting to order.

一般来说，会议记录以会议主持人的姓名开始，也可能包括会议地点、日期、在场人员名单以及主席宣布会议开始的时间。

Since the primary function of minutes is to record the decisions made, all official decisions must be included. If a formal motion is proposed, seconded, passed, or not, then this is recorded. The voting tally may also be included.

由于会议记录的主要功能是记录所做的决定，因此会议记录必须包括所有正式决定。所有会上提出的议案、与会者复议议案，议案通过或不通过都必须记录。投票计数也可以包括在内。

Minutes are sometimes submitted by the secretary at a subsequent meeting for review. Usually, one of the first items in an agenda for a meeting is the reading and approval of the minutes from the previous meeting.

会议记录有时由秘书在随后的会议上提交审查。通常，会议议程中的第一项内容之一是阅读和批准上次会议的会议记录。

If the members of the group agree that the written minutes reflect what happened at the previous meeting, then they are approved, and the fact of their approval is recorded in the minutes of the current meeting. If there are significant errors or omissions, then the minutes may be redrafted and submitted again at a later date. Minor changes may be made immediately using the normal amendment procedures, and the amended minutes may be approved "as amended".

如果小组成员同意会议纪要反映了在前一次会议上发生的事情，那么它们将被批准，并且其批准的事实也会被记录在本次会议的会议记录中。如果存在重大错误或遗漏，则可以重新起草会议记录并在稍后日期再次提交。可以使用正常的修改程序立即作出小的更改，修改后的会议记录可以"经修改"批准。

The commonly used format and useful expressions of meeting minutes are as follows:

常用的会议记录的格式和有用句型如下：

[Company/Department Name]
Meeting Minutes
[Date]

I. Call to order

The CEO **called the meeting to order** at 9:00 a.m. at Conference Room 1805.

The Chairman **called the meeting to order** at 2:05 p.m.

II. Roll call

The company secretary **conducted a roll call**.

III. Approval of minutes from the last meeting

The Chair **approved the minutes** from the October 10, 2014 Board meeting.

The minutes of the last meeting was approved by the Board of Directors.

IV. Open issues

The project of ... is yet to be further discussed in the meeting.

Please get ready for further discussion of the ... Program.

V. New business

... provided the Board with the information regarding support to the new project.

Board members move to discuss the prospect of the new ... Project.

VI. Adjournment

Mr. ... thanked everyone present and ended the meeting.

The Board meeting adjourned at 4:15 p.m.

Activity 11: Please fill in the blanks of the meeting minutes with the words or phrases in the table. 请将下表中的词或短语填入会议记录的适当位置。

A. Approval	B. Adjournment	C. Open issues	D. in the Conference Room
E. The Foster Lash Company, Inc		F. Absent	G. New business

1

Capital Improvement Committee Meeting Minutes
October 8, 2014

I. Call to order

The weekly meeting of the Capital Improvement Committee of the Foster Lash Company was called to order at 11 a.m. _____2_____ by Mr. Stuart.

II. Roll call

The company secretary conducted a roll call.

  Present: Mike Negron, Sheila Glun, Ellen Frankin

  _____3_____: Fred Hoffman, Gina Marino

III. _____4_____ **of minutes from the last meeting**

The minutes of the meeting of October 1 were read by Mr. Negron and approved.

IV. _____5_____

The main discussion of the meeting concerned major equipment that should be purchased by the end of the year. Among the proposals were these:

Mr. Woo presented information regarding three varieties of office copying machines. On the basis of her cost-benefit analysis and relative performance statistics, it was decided, by majority vote, to recommend the purchase of a CBM X-12 copier.

Mr. Browne presented a request from the secretarial staff for new typewriters. Several secretaries have complained of major and frequent breakdowns of their old machines. Mr. Franklin and Mr.

Browne are to further investigate the need for new typewriters and prepare a cost comparison of new equipment versus repairs.

V. _____6_____

The committee will discuss the advisability of upgrading account executives' personal computers. The report will be presented by Sheila Glun at the next meeting, to be held on October 15,2014, at 11 a.m. in the Conference Room.

VI. _____7_____

Mr. Stuart thanked everyone present and ended the meeting.

The meeting adjourned at 11:45 a.m.

Respectfully submitted,
Ellen Franklin
Secretary

Speaking Skills 交际技巧

巧用程度副词

副词一直以来被很多同学忽略并且认为简单易懂，但是在听说读写中，副词对句子语义的表达往往起着重要作用，在口语交际中，程度副词的使用尤其频繁。

那么程度副词其魅力到底在哪里，什么样的词我们将其归类于程度副词，这样的词在英语中又该如何使用呢？今天我们就在这里详细地给大家解读下程度副词这个神奇的存在。英语中程度副词可以分为两大类：加强情感语气类程度副词和限定动词方式类程度副词。

第一类在口语中是屡见不鲜的，很多同学可能不经意间已经使用过类似的语言了。比如说："That's extremely brilliant." 在这个句子中 extremely 就是很典型的程度副词用来加强语气的，说话人表达的不是很美妙，是极其美妙、美妙得无与伦比。这类程度副词主要包括表示完全、彻底的形容词构成的副词，比如：totally, definitely, exactly, excessively, completely 和 extensively, perfectly, absolutely, undoubtedly, 等等。当人们想表达××后果严重，或许会用 very，但是 completely 可用来表达更深的无奈和郁闷。例如，

A: She is the best teacher I have ever met. 她是我遇到过的最好的老师。

B: Totally. 完全正确。

第二类程度副词在日常应用中也层出不穷，比如 "He took his toys extremely seriously and would like to play with them endlessly." 在这个句子中使用了 extremely, endlessly 两个程度副词，表达动作的发生方式。类似动词主要需要大家关注的是一些形容词+ly 发生词义的变化，比如，dreadful 表示恐怖的，但是 dreadfully 表示的是极其。这样的动词用在我们同学的写作中肯定会吸引读者的眼球，使大家对作者的写作态度和立场有更深入的了解。

类似的程度副词还有 especially, drastically, awfully, particularly, really, 等等。

After practicing with my foreign English teacher, my oral English has been improved drastically. 通过跟外教的练习，我的英语口语有了很大提高。

Activity 12: Please choose the most appropriate adverb to fill in the blanks. 请选择最合适的副词填空。(可多选)

| A. absolutely | B. completely | C. definitely | D. particularly |

1. Everything about the country seemed _____ different from what I'd experienced before.
2. Most of the travelers, _____ old people, do not like long journey.
3. _____, it is necessary for people to have a solid understanding of their country's history.
4. If you do not want to do that, reject it _____.

Activity 13: Please translate these sentences into English, with right adverb of degree we have learned. 请将下列句子翻译成英语，注意使用学过的程度副词。

1. 这事还没有明确解决。
_____

2. 我绝对同意你的观点。
_____

3. 这些日子我们特别忙。
_____

4. 他用所有的积蓄买了那个他特别喜欢的房子。
_____

5. 毫无疑问，他是中国的骄傲。
_____

Self-evaluation

Items	Yes	No
I have acquired all the key words in the Mission Objective		
I have acquired all the sentence patterns in the Mission Objective		
I know the format of Meeting Minutes		

Unit 3

Job Fair 双选会

 Warming-up 课程准备

Mission Objective 任务目标

1. 掌握笔试、面试、招聘宣讲、职位、福利、面试官、雇主、应聘者、职业风险、发展前景等英语表达，了解常见的词语后缀。

2. 会用英语表述双选会的流程、会表达自己的喜好、描述自己的工作经历和发展方向，会根据情境选用恰当的连接词使口语表达更有层次。

3. 会填写入职登记表。

Discussion 讨论

1. What should we prepare before interview?
2. Will you be nervous during an interview? How to overcome it?

Words and phrases for assistance:

面试: documents, resume, information, post, position, suitable, preparation, dress code, formal, calm, confident, deep breath, fully prepared, eye contact, arrive ahead of time

Database 知识库

Job Interview Tips for Job Seekers

1. Conduct Research on the Employer, Interviewer, and Job Opportunity

Success in a job interview starts with a solid foundation of knowledge on the job seeker's part. You should understand the employer, the requirements of the job. The more research you conduct, the more you'll understand the employer, and the better you'll be able to answer interview questions. Scour the organization's website and other materials, and ask questions about the company.

2. Review Common Interview Questions and Prepare Your Responses

Another key to interview success is preparing responses to expected interview questions. First, ask the hiring manager as to the type of interview to expect. Will it be one-on-one or in a group? Will it be with one person, or will you meet several members of the organization? Your goal is to try to determine what you'll be asked and to compose detailed yet concise responses that focus on specific examples and accomplishments.

3. Dress for Success

Plan out a wardrobe that fits the organization and its culture, striving for the most professional appearance you can accomplish. Remember that it's always better to be overdressed than under. Keep accessories and jewelry to a minimum. Try not to smoke or eat right before the interview.

4. Arrive on Time, Relaxed and Prepared for the Interview

There is no excuse ever for arriving late to an interview. Strive to arrive about 15 minutes before your scheduled interview to complete additional paperwork and allow yourself time to get settled. Arriving a bit early is also a chance to observe the dynamics of the workplace.

The day before the interview, pack up extra copies of your resume. Finally, remember to pack several pens and a pad of paper to jot notes. Finally, as you get to the offices, shut off your cell phone.

5. Make Good First Impressions

A cardinal rule of interviewing is to be polite and offer warm greetings to everyone you meet from the parking attendant to the receptionist to the hiring manager. Employers often are curious how job applicants treat staff members and your job offer could easily be derailed if you're rude or arrogant to any of the staff. When it's time for the interview, keep in mind that first impressions "the impression interviewers get in the first few seconds of meeting you" can make or break an interview.

Remember that having a positive attitude and expressing enthusiasm for the job and employer are vital in the initial stages of the interview; studies show that hiring managers make critical decisions about job applicants in the first 20 minutes of the interview.

6. Thank Interviewer(s) in Person, or by E-mail

Common courtesy and politeness go far in interviewing; thus, the importance of thanking each person who interviews you should come as no surprise. Start the process while at the interview, thanking each person who interviewed you before you leave. Writing thank-you e-mails or notes shortly after the interview will not get you the job offer, but doing so will certainly give you an edge over any of the other applicants who didn't bother to send thank-you letters.

Activity 1: Fill in the blanks according to the passage. 根据短文内容填空。

1. 面试 _____
2. 面试官 _____
3. 雇主 _____
4. 候选人 _____
5. 第一印象 _____
6. 热情的问候 _____
7. 接待员 _____
8. 感谢信 _____
9. 网站 _____
10. 简历 _____
11. 副本 _____
12. 工作机会 _____
13. 积极的态度 _____

14. 求职者 _____

15. 录用通知 _____

Activity 2: Group work. Please try to make a short passage using the above-mentioned words. 小组活动。请尝试用上面的词语写一段话。

✸ Oral Practice 口语训练

人物涉及 (Characters):

C: Mr. Chen (Director of Sino-German Technical College) 中德技术学院主任

W: Wang Nan (Freshman Majoring in Industrial Process Automation Technology) 工业过程自动化技术专业新生

Z: Zhang Yang (Apprentice of FESTO) FESTO 签约学徒

情景解说 (Situations):

Wang Nan(W) 听说中德技术学院即将召开双选会，所以去 Mr. Chen(C) 的办公室，围绕选拔流程，以及选拔中该注意的问题进行了咨询。

在企业宣讲之后，Wang Nan(W) 和 Zhang Yang(Z) 讨论自己中意的公司和对自己职业生涯的规划。

最后，Wang Nan 顺利地通过了笔试，接受企业培训经理 Mr. Li(L) 面试。

以下对话围绕这三个情景展开。

Dialogue 1 Consulting the Teacher about the Job Fair 向老师咨询双选会

W: Hi, Mr. Chen. I heard that we would have a **job fair**① soon. Could you tell me more about it? C: Hi, Wang Nan. That's a good question. It is surely② an important choice to everyone of you.	双选会

030

W: Yes, I feel a little bit **nervous actually**.	紧张，事实上
C: Take it easy ③. The job fair will be held next week. All the companies in the program will give **recruitment talk** and introduce **posts** they will provide, the company, their products, the **welfare** and your **development prospects**.	招聘宣讲，职位 福利，发展前景
W: That's good for us to have a **comprehensive** understanding of the companies.	全面的
C: Yes. You can make **consultations** then and choose the companies you like. After **a paper test** and a **face-to-face interview**, the corporations will choose satisfied freshmen and **sign them**.	咨询 笔试，面试 与他们签约
W: Do I need any preparations **in advance**?	提前
C: Don't worry about the test. It will not focus on professional knowledge, but your **creative thinking**, **spatial thinking** and **imagery thinking**.	创造性思维，空间思维，形象思维
W: What about the interview?	
C: It is **beneficial** if you are well prepared. You can show your **previous achievements** ④, and your **works** if you have. Also some popular freshmen can receive more than one **offer**, so they can make a decision. ⑤	有帮助，之前的 成绩，作品 录用通知
W: Now I have a clear picture of what will happen. Thank you, Sir.	
C: Wish you a big success! ⑥	

Note:

① fair 在本句中是集市、展览会的意思，job fair 是招聘会，中国义乌有 China Yiwu International Commodities Fair（中国义乌国际小商品博览会）。另外，fair 还有公平的意思，反义词是 unfair。例如，fair play（公平竞争）是西方人特别重视的原则。

② surely 的意思是无疑地、肯定地，在句子中加强语气。例如，You surely haven't forgotten Mr. White? 你肯定还没忘记怀特先生吧？

③ take it easy 多用于别人对某事感到不安的情况下，意思是轻松点，别紧张，别着急。

④ 在本对话中出现了较多的带后缀的单词。后缀是一个单词的词性的重要标记。掌握常见的后缀是扩大词汇量的重要手段。recruitment, development, achievement 中的 –ment 是名词后缀，常用于表示行为、状态、过程、手段及其结果，例如，treatment, movement, judgment, punishment, argument 等；consultation 中的 –tion,–ation 也是名词后缀，表示"行为的过程、结果、状况"，例如，action, solution, conclusion 等；comprehensive, creative 中的 –ive 是形容词后缀，带有"属性、倾向、相关"的含义，例如，active, impressive, decisive 等；beneficial 中的后缀 –ial 表示"有……的"，例如，provincial, commercial 等。

⑤ Also some popular freshmen can receive more than one offer, so they can make a decision. 另外，受欢迎的新生可能收到不止一家的工作邀请，他们可以自己做决定。Offer 的全称是 offer letter，指的是"录用信"或"录取通知"。offer 指的是外企或国外学校发的表达自己愿意录用或录取的一封正式的信件。

⑥ Wish you a big success! 是祝你成功的意思。也可以说 May you succeed!。

Activity 3: Change the following words into the part of speech required by adding suffixes. 将下列单词加上适当的后缀改成要求的词性。

1) *into nouns*

　　equip ＿＿＿＿＿＿＿＿＿＿＿＿＿＿　　　　　　govern ＿＿＿＿＿＿＿＿＿＿＿＿＿＿

state _____   move _____
advertise _____

2) into adjectives

sense _____   product _____
expense _____   mass _____
act _____

Activity 4: Oral Practice 口语练习

Role play this dialogue with your partner.

Activity 5: Please summarize the procedures of the job fair according to Dialogue 1. 请根据对话 1 总结双选会的几个步骤。

Step 1: _____
Step 2: _____
Step 3: _____

Dialogue 2   Discussing the Future Job with Classmates 与同学讨论未来的工作

W: Zhang Yang, _____ 1 _____ ?	
Z: Because it is the **global** leader in **bionic technology** and **pneumatic automation field**, which is fantastic ① ! What's more ② , FESTO-Jinan **enjoys a very good reputation** and offers welfare to the apprentices like free lunch and free accommodation ③. _____ 2 _____ ?	全球的，仿生学，气动，自动化，领域 享有良好声誉
W: I prefer ④ companies in the auto field. I think **part suppliers** for auto industries are **promising**. So Voss is my **first choice**.	零件供应商 有前途的，首选
Z: Actually, all the companies have their own **characteristics.** _____ 3 _____.	特色
W: That's **reasonable**. Do you know anything about the **occupational risks**?	有道理，职业风险
Z: The noise in the workplace will cause ear **discomfort** and long-time standing hurts the knees, in addition ②, you need to work in three **shifts**.	不适 轮班
W: After **signing the contract**, _____ 4 _____.	签约
Z: Yes. In the second year, we will begin the **on-job training** in the signed corporation. Just think about your future carefully and make an **intelligent** choice.	岗位培训 明智的
W: Thanks for your suggestion, Zhang Yang.	

Note：

① fantastic 在口语中经常出现的，意思就跟 wonderful, gorgeous 一样，都表示 "很棒，好极了"。例如，She's doing a fantastic job. 她干得很漂亮。Wow! You look fantastic! 哇，你看起来真精神！

② what's more 和下文中的 in addition 都表示递进关系，是 "还有……" 的意思，既可以放句首，也可以放句中。例如，He is friendly to us. What's more, he is clever. 他对我们很友好，而且他很聪明。In addition, we should smile and appear friendly. 此外，我们应该微笑，显得友好。

③ FESTO-Jinan enjoys a very good reputation and offers welfare to the apprentices like free lunch and free accommodation. 济南费斯托公司的声誉很好，并且他们给学徒提供了很多的福利，比如免费的午餐和免费的住宿。enjoy 在本句中是 "享有，拥有" 的意思，例如 enjoy paid holiday 享受带薪休假，enjoy good health 身体健康。另外，enjoy 还可以表示喜欢，如 enjoy the film, enjoy singing. 加反身代词表示"过得愉快，玩得高兴"，如 enjoy yourself, enjoy myself。

④ prefer 的用法：

prefer + 名词或动名词表示"宁愿""更喜欢"。例如：He comes from Shanghai, so he prefers rice. 他是上海人，因此更喜欢吃米饭。I prefer going by bike. 我宁愿骑单车去。I prefer the white one. 我喜欢那个白的。

prefer A to B (AB 皆为名词) 时，表示与 B 相比更喜欢 A。例如：I prefer the blue ball to the black ball. 我更喜欢蓝色球而不喜欢黑色球。

Activity 6: Choose the best choice from A to D to finish the dialogue. 将 A–D 填入对话中完成对话。

| A. We will make progress as long as⑤ we work hard | B. why did you choose FESTO |
| C. Which company will you choose | D. we will become signed apprentices |

⑤ as long as，只要。例如，As long as you have determination, you can do anything. 只要有决心，你可以做任何事情。

Activity 7: Make sentences using "as long as". 用"as long as"造句。

1. There is life.   There is hope.

_____

2. You can go out.   You promise to be back before 11 o'clock.

_____

3. You'll succeed.   You work hard.

_____

4. You need me.   I'll stay.

_____

5. Electric current flows through a wire.   There is a potential difference.

_____

Activity 8: Oral Practice 口语练习

Role play this dialogue with your partner.

Dialogue 3   During the Interview 进行面试

L: Good afternoon, Wang Nan. Please **take a seat**.	请坐
W: Good afternoon.	
L: Well, Wang Nan, I know from your **Application Form** that you come from Qingdao. Are you willing to① work far from your home, like in Jinan?	申请表
W: Yes. No problem if the offer is really attractive.	
L: Can you tell me more about your working experience?	
W: I've worked in the **Marketing Department** of a sportswear company, as a **salesman**. I've learned a lot of working experience in the sales and marketing, especially how to **communicate with** the customers.②	市场部 销售员 交流

L: I see. Why would you like to join in this program? W: I think I can learn practical skills, because the students in this program spend more practicing time than others. L: What would you like to do in the future? W: I am interested in Industrial Process Automation Technology, so I'd like to be an **engineer** in this **field**.③ L: Well, you are quite **ambitious** and **enthusiastic**. Can you talk about other advantages? W: I think I'm **ingenious** and hard-working. L: What would you say about your **weakness**? W: Ah, that's a difficult question to answer. I sometimes can be **impatient,** so interest is an important factor to my future job. L: Thank you for your **attendance** and I'll **inform** you as soon as possible.④ W: Thank you and I am **looking forward to** hearing from you.⑤	工程师，领域 有抱负的，有热情的 心灵手巧的 弱点 没有耐心的 出席，通知 期待

Note:

① willing to 是"乐意"的意思。例如，I am willing to do anything for you. 我很愿意帮你做任何事情。

② I've learned a lot of working experience in the sales and marketing, especially how to communicate with the customers. 我在销售和市场方面学习了许多经验，尤其是如何与客户交流方面。

③ I'd like 是"I would like to ..."的缩写，是一种客气的表达自己想法的说法。例如，I'd like to invite you to my birthday party this weekend. 我想邀请你周末来参加我的生日会。

④ Thank you for your attendance and I'll inform you as soon as possible. 感谢您的出席，我会尽快通知您。

⑤ look forward to 表示"期待；盼望"，这里的 to 是介词，后面接名词或者动词的 ing 形式。比如：He has been looking forward to going to England for a long time. 好久以来他一直盼望去英国。

Activity 9: Fill in the blanks with the right form of the phrases from the box to complete the following sentences. 用所给短语的正确形式填空完成句子。

consult with	in advance	focus on	communicate with	look forward to

1. I wanted to _____ my teacher about the program.
2. _____ the task until it is done.
3. He is _____ working with the new manager.
4. I can _____ with foreigners easily because of good English.
5. Remember to book _____, otherwise it will be difficult to get accommodation.

Activity 10: Match the phrases with their Chinese equivalents. 将英语单词与对应的汉语连线。

1. ambitious   A. 不耐烦的
2. enthusiastic   B. 有雄心的、有抱负的
3. impatient   C. 热情的
4. creative   D. 创造性的
5. professional   E. 双的，二重的
6. dual   F. 专业的，职业的

Activity 11: Translate the following sentences. 翻译下列句子。

1. Jim has been working hard and looks forward to spending his vacation.

2. Frank was looking forward to that evening's date.

3. He is looking forward to going to England.

4. 他盼望着和新来的经理一起工作。

5. 我们盼望再次见到您！

Activity 12: Oral Practice. 口语练习。
Try to describe your working advantages to your classmates.

Reading and Writing 读写拓展

Reading 阅读

STIHL is headquartered in Wieblingen, which is a city in Baden-Württemberg, south-west Germany. Wieblingen has a population of 52,948, with an area of 42.76 square kilometers. Wieblingen belongs to the administrative district of Stuttgart, the county is Remus –Moore County, 230 meters above sea level.

STIHL (ANDREAS STIHL Power Tools (Qingdao) Co., Ltd.), as the global market leader of chain saw and power tools, its power saw has been eternally No. 1 in the world.

 power tools 电动工具；有动力的工具
 chain saw 链锯；小型机器锯，油锯
 power saw [木] 动力锯

Activity 13: Read the following passage and finish the questions below. 读短文回答问题。

  The purpose of an interview is to find out if your goals and the goals of an organization are *compatible*. Other goals of the interview are: to answer questions successfully, obtain any additional information needed to make a decision, establish a positive relationship, show confidence, and to sell yourself. Based on these goals, place yourself in the role of the interviewer and develop anticipated questions and answers to three categories: company data, personal data, and specific job data.

To develop confidence, adequately prepare for the interview. Focus on how you can best serve the organization to which you are applying.

Since the interview will center on you, proper self-management process is divided into four stages: the before stage, the greeting stage, the consultation stage, and the departure stage. The before stage includes writing a confirmation letter, concentrating on appearance and nonverbal communication, developing your portfolio, anticipating questions with positive responses, and arriving early. The greeting stage includes greeting everyone courteously, using waiting-room smarts, using your time wisely, and applying proper protocol when meeting the interviewer. The consultation stage includes responsiveness and enthusiasm, knowing when to interject key points, showing sincerity, highlighting your strengths, and listening intently. The departure stage includes leaving on a positive note, expressing appreciation, expressing interest, leaving promptly, and making notes immediately after departure.

Following the interview, write thank-you letters to each person who interviewed you and to those who helped you get the interview. When invited for a second interview, go prepared by using your notes and feedback from the interview to zero in on what the company wants. If the company doesn't respond in two weeks, call back or write a follow-up letter. You may get turned down. If so, try to find out why as a means of self-improvement.

Following a job offer, take a few days to consider all elements and then call or write a letter either accepting or declining the offer—whichever is appropriate. If you accept and you are presently employed, write an effective letter of resignation, departing on a positive note.

1. The word "compatible" in the first sentence probably means _____.

   A. in agreement   B. in conflict
   C. complementary   D. practicable

2. At which stage should you emphasize your qualifications for the job?

   A. The before stage.   B. The greeting stage.
   C. The consultation stage. D. The departure stage.

3. If you are given a second interview, it is most important for you to _____.

   A. write a thank-you letter to each person who interviewed you last time
   B. find out exactly what the company wants of you
   C. learn from the last interview and improve yourself
   D. consider all the elements that are important for the job

4. The passage is mainly concerned with _____.

   A. how to manage an interview
   B. how to apply for a job vacancy
   C. how an applicant should behave during an interview
   D. how to make your private goal compatible with those of an organization

Writing 写作

Employee Entry Registration Form 员工入职登记表

个人信息 Personal Information				
姓名 Name				
性别 Gender			国籍 Nationality	
出生日期 Date of Birth			出生地 Place of Birth	
婚姻状况 Marital Status	☐ 已婚 Married	☐ 未婚 Single	☐ 离异 Divorced	
家庭电话 Home Tel				
手机 Mobil Phone				
邮箱 E-mail Address				
紧急联络人 Emergency contact person				
姓名 Name	关系 Relationship	联系地址及邮编 Address& Zip Code	电话 Phone Number	电子邮箱 E-mail Address

健康状况 Health Condition	身高 Height(　　)　　体重 Weight(　　)
	视力 Vision　　(　)良好 Good　　(　)矫正 Assist
	听力 Hearing　(　)良好 Good　　(　)矫正 Assist
	是否曾被认定为工伤或职业病或持有残疾人证明：填写"是"或"否"（　　） Whether identified work injury, occupational disease or hold certificate of disability: Please fill in "Yes" or "no"
	是否被劳动能力鉴定委员会鉴定为具有伤残等级以及何级伤残：填写"是"或"否"以及伤残等级（　）（　　） Whether identified by labor appraisal committee as having disability and its grade: Please fill in "Yes" or "no" and the degree of disability
	是否从事过井下、高空、高温、特别繁重体力劳动以及有毒有害工种：填写"是"或"否"（　　） Whether engaged in underground, high altitude, high temperature, extremely heavy manual labor, as well as poisonous and harmful work: Please fill in "yes" or "no"
	是否有传染性疾病以及何疾病：填写"是"或"否"以及何疾病（　　）（　　） Whether having infectious disease : Please fill in "yes" or "no" and which disease.
	最近6个月内所接受的医学治疗与医学检查： Medical treatment and examination within the latest 6 months

教育背景 Education Background	
起止日期 From – To	
学校名称 Name of the College/University	
专业 Major	
学位 Degree	
起止日期 From – To	
学校名称 Name of the College/University	
专业 Major	
学位 Degree	
专业资格与培训 Professional Qualification & Training	

前用人单位信息 The last company information	离职时间 Resignation date		离职原因 Resignation reason	
	是否与前用人单位约定了保密协议与竞业限制条款：填写"是"或"否"（　　） Whether signed confidentiality agreement and non-competition clause with former company: Please fill in "yes" or "no"			
	是否与前用人单位有未尽的法律事宜：填写"是"或"否"（　　） Whether having legal matters not over yet with former company: Please fill in "yes" or "no"			

参加工作时间 Date of attending job		累计工作时间 Total working time	
是否已经休了本年度的年休假：填写"是"或"否"（　　） Whether you have already enjoyed annual leaves this year: Please fill in "yes" or "no"		是否曾经或正在追究与承担过刑事责任：填写"是"或"否"（　　） Whether you have been involved in any criminal issues: Please fill in "yes" or "no"	

应聘信息来源 Source of recruitment information		是否在本公司工作过：填写"是"或"否"（　　） Whether worked in our company: Please fill in "yes" or "no"	
入职部门 Entry department		入职职位 Job Title	
		入职时间 Hire date	

员工声明 Statement	1. 员工确认，公司已如实告知工作内容、工作地点、工作条件、职业危害、安全生产状况、劳动报酬以及员工要求了解的情况。 I confirmed that Company has truthfully informed working content, working place, working conditions, occupational hazards, production safety conditions, labor remuneration and other information I want to know.

续表

员工声明 Statement	2. 员工在本表提供的个人信息、学历证明、资格证明、身份证明、工作经历等个人资料均真实，员工充分了解上述资料的真实性是双方订立劳动合同的前提条件，如有弄虚作假或隐瞒的情况，属于严重违反公司规章制度，同意公司有权解除劳动合同或对劳动合同做无效认定处理，公司因此遭受的损失，员工有对此赔偿的义务。 I promise all the information registered in this form is true including—The personal information, education certificates, qualification certificates, proof of identification, working experiences and so on. I fully understand the importance of above information's authenticity which is the premise of labor contract. If there is any cheating and fake information here, I agree that company can terminate our labor contract without any compensation and I will afford the loss brought to company. 3. 员工确认，本表所填写的通信地址为邮寄送达地址，公司向该通信地址寄送的文件或物品，如果发生收件人拒绝签收或其他无法送达的情形的，员工同意，从公司寄出之日起视为公司已经送达。 I confirmed, the mailing address I filled in this form is correct and can be delivered by express. I agree that it should be regarded as I have already received all the documents or goods sent by Company to this address even it happens that they are refused or cannot be delivered. 员工签名： 日期： Signature: Date:
单位填写 Filled by company	试用期限 Probation period 试用期工资 Probation salary 正式期工资 Regular salary
员工确认 Confirmed by employee	本人对入职登记表上登记的全部内容皆已知晓并保证我所提供以及填写的资料均属实。 I have already known and understood all contents in this Entry Registration Form, and ensure all the information provided by me is real. 签名及日期 Signature &Date:

Speaking Skills 交际技巧

巧用连接词

连接词即表示起承转合的词汇，能够将两个或以上的句子连接在一起并显示关联成分之间的逻辑关系，在英语中起连接架构的作用。恰当地使用连接词不仅能够让语言连贯、突出文章的层次感和逻辑性，而且能更好地表达自己的思想，实现有效交流的目标。常用的连接词有以下几类：

1. 表示逻辑上的先后顺序的连接词

1）首先：first，firstly，in the first place，to begin with 等。

2）其次：secondly，in the second place 等。

3）最后，最重要的是：at last，finally，most importantly 等。

4）最后但并非最不重要的（一点）是：last but not least。

当列举几个并列的原因、行动时，可以使用这一类连接词，让听者更容易获得信息。例如，*I have many plans for this weekend. **First**, I have to do my homework so that I can finish it on time. **Second**, I want to watch TV because there are some movies I want to enjoy. **Third**, I plan to play football with my friends. **Last**, I want to have an excellent sleep.*

2. 表示递进关系的连接词

1）也，而且，还：also，too，besides

2）此外：in addition，apart from，furthermore，what's more，moreover

3）不仅……而且……：not only... but also...

4）既……又……，也：both... and...，as well as

有时说话人想要表达两个或以上并列的意思，但其中需要强调某一点，这时，可以选择不同含义的连接词使用，突出语言表达中的重点。例如，*You need money and time. **In addition**, you need diligence.* 你需要时间和金钱。除此之外，你还需要努力。

3. 表示因果关系的连接词

1）(后接表原因的从句) 因为：because，for，because of...，owing to...

2）因此，所以：thus，hence，therefore

3）那么：then

4）结果（是）：as a result，so that（后接表结果的从句）

因果关系是最常见的逻辑关系之一。恰当地使用表示因果关系的连接词，会使逻辑关系更清晰，让听者更容易抓住重点。*The cost of transport is a major expense for an industry. **Hence** factory location is an important consideration.* 运输成本是工厂重要的支出。因此，工厂选址是重要的考虑因素。

4. 表示举例的连接词

例证，也就是举例说明，恰当的举例能够帮助谈话者用更多的细节进一步解释自己的话，使自己的语言更准确，达到沟通顺畅的效果。

通常例证都出现在所要表达的观点之后。例如，*I have taken part-time jobs in my spare time, **for example**, I have worked in the Marketing Department of a sportswear Company.*

除了 for example，以下词汇和短语也能用于举例说明。例如，***To illustrate**, Figure 4 shows the machining process of a CNC milling machine.* 为了加以说明，图4演示了数控铣床的加工工艺。

1）也就是说：namely，that is，that is to say

2）例如：for example，for instance，such as，to illustrate

3）以……为例（来说）：take ... as an example

4）等等，其他类似的：... and things like that，something like that

*There are three kinds of fits, **namely**, clearance, transition and interference.* 配合有三种，它们是间隙配合、过渡配合和过盈配合。

5. 表示转折的连接词

表示转折时，除了常用的 but，however 等连词外，还可以根据不同的语境，用 although 等连接词来表示语义的转变，会让你的口语脱颖而出！例如，*They fulfilled the plan, although they worked under the unfavourable conditions.* 尽管他们在不利的条件下工作，他们仍完成了计划。

1) 然而 (and)：yet，while，whereas，nevertheless
2) 尽管，虽然：though，although
3) 相反：on the contrary
4) 与……形成对比：in contrast with/to
5) 代替，而不是：instead，instead of
6) 毕竟：after all

Activity 14: Please fill in the blanks using the given linking words. 用所给连接词填空。

| as a result | Apart from | as well as | furthermore | Most importantly |

1. _____ singing a song, he told a joke.
2. He didn't work hard, _____, he failed his exam.
3. _____, she had the courage to try.
4. We are repairing the roof _____ painting the walls.
5. The house is beautiful. _____, it's in a great location.

Activity 15: Please combine the following sentences using linking words and phrases. 请用恰当的连接词或词组连接下列各组句子。

1. The ground is wet _____ it rained last night.
2. You can't tell many things just from the appearance, _____ the past, hobbies, eating habits and so on.
3. He taught us _____ a subject, _____ how to carry on learning about the subject after class.
4. Why not call some friends and have a picnic _____ eating out?
5. _____ giving a general introduction to computer, the course also provides practical experience.

Self-evaluation

Items	Yes	No
I have acquired all the key words in the Mission Objective		
I have acquired all the sentence patterns in the Mission Objective		
I can fill in the Employee Entry Registration Form		

Unit 4
Working Environment 工作环境

❋ Warming-up 课程准备

Mission Objective 任务目标

1. 掌握公司各部门、手动加工车间常用工具、刀具、量具，机床加工车间常用设备（包括车床、铣床、磨床、加工中心等），以及工程应用软件的英文表达。

2. 会用英语说明工作职责，简单介绍手动加工项目的训练内容和机床加工项目的学习内容，会用英语指路，会使用举例子的连接词。

3. 会填写申请表，用英语撰写申请信。

Discussion 讨论

1. What is your impression of the German plants?
2. Are they the same with your imagination?

Words and phrases for assistance:

空间和场所：clean, tidy, organized, modern, spacious, bright, advanced, high-tech, machines

Database 知识库

Duties of Departments

Production Department

- Responsible for the company's production management
- Complete the company's production tasks on time
- Analyze and solve major production technical problems and quality problems in the production process
- Guide the production workshop to do the "6S" management work on the production site

Purchasing Department

- Select suppliers
- Prepare purchase orders
- Manage customer relationship and deal with suppliers

Logistics Department

- Take/receive and manage sales orders

- Follow the situation of the goods arriving
- Transfer the notice paper of disqualified goods
- Control the invoice of purchased goods
- Manage inventories, deliver/dispatch goods

Research and Development Department
- To be engaged in new products development and management
- According to the market situation, make new technology strategy
- Research industry technology development trend, and explore new projects and new product
- Research to understand the product market dynamic, submit product development proposals
- Prepare the new product development plan and organize the implementation
- Make new products and market promotion of trial production

Personnel/Human Resources Department
- Recruit new staff
- Interview applicants
- Train staff
- Fire staff
- Deal with relationships between management and work force

Manager Office
- Make /Answer telephone calls
- Deal with files and correspondence
- Write memos and reports
- Word processing
- Receive guests and visitors
- Arrange meetings
- Make appointments

Customer Service Department
- Deal with customers' problems
- Accept repairs and deal with customers' complaints
- Communicate timely with customers
- Offer a comprehensive and convenient after-care service
- Offer a full professional after-sales service
- Undertake installation and testing tasks
- Offer training to users' technical staff

Manager
- Be responsible for the main activities of the company
- Manufacture and sale
- Make decisions
- Make plans
- Handle difficult situations at work

- Deal with problems and complaints

Activity 1: Fill in the blanks according to the passage. 根据短文填空。

1. The _____ Department is responsible for developing new products.
2. The _____ Department connects with suppliers.
3. The _____ Department connects with customers.
4. The R&D pays its attention to _____, _____.
5. The _____ Department focuses on staff.
6. The _____ Department deals with "6S" management on the production site.

Activity 2: Match the department with their Chinese equivalents. 将各部门名称与其汉语连线。

1. Production Department   A. 客服部
2. Purchasing Department   B. 人力资源部
3. Logistics Department   C. 采购部
4. Research and Development Department   D. 物流部
5. Human Resources Department   E. 生产部
6. Manager Office   F. 研发部
7. Customer Service Department   G. 经理办公室

Activity 3: Oral Practice. Please try to list your duties when you did the above-mentioned group work. 口语练习。请尝试描述你在小组活动中承担的职责。

I'm responsible for _____.
I pay attention to _____.
I focus on _____.
I prepare/offer _____.

Oral Practice 口语训练

人物涉及 (Characters):

G: Gao Jian (Freshman Majoring in CNC Technology) 数控技术专业新生

C: Mr. Chen (Director of Sino-German Technical College) 中德技术学院陈主任

W: Wang Nan (Freshman Majoring in Industrial Process Automation Technology) 工业过程自动化技术专业新生

Q: Mr. Qi (A Team Leader Working in Logistics Department) 物流部班组长

Z: Zhang Yang (Signed Apprentices of FESTO) 费斯托签约学徒

情景解说 (Situations):

中德技术学院 Mr. Chen(C) 带领 Wang Nan (W)、Gao Jian(G) 等新生参观手动加工车间和机床加工车间，了解学习环境和学习内容。

签约学徒 Zhang Yang (Z) 去企业实习，找不到供应室在哪里，向路过的班长 Mr. Qi(Q) 问路。

Dialogue 1　Workshop of Manual Machining 手动加工车间

C: This is "Workshop of Manual Machining", which is **the same as** German training workshop.①	与……相同
G: The **surfaces of workbenches** are all made of wood. Is this a classroom? Why do we set a classroom in the workshop?	工作台台面
C: The classroom is used for theory learning. You can **put the knowledge into practice directly** in the workshop.	将知识付诸实践 直接
G: What can we learn in this workshop?	
C: You can learn the knowledge and skills of **manual machining**—the work as a **fitter**, such as the types of tools, **cutting tools** and **inspecting tools**②. Besides, you will learn the **safety knowledge**. Moreover, you can do the **process analysis** of some simple **parts**.③	手动加工 钳工，刀具，量具 安全知识，工艺分析 零件
G: What's that?	
C: It's called **bench drilling machine**. The **operation** skills of bench drilling machine is essential for CNC Technology majors.④	台式钻床，操作

Note:

① This is "Workshop of Manual Machining", which is the same as German training workshop. 这是"手动加工车间"。这个车间跟德国的培训车间是一样的。

② inspecting tools 也叫做 measuring tools，是量具的意思。中德实训车间常用的量具包括游标卡尺 (vernier caliper)、高度尺 (height caliper)、千分尺 (micrometer)、刀口角尺 (try square)、钢尺 (steel ruler)、塞尺 (feeler gauge) 等。

③ You can learn the knowledge and skills of manual machining—the work as a fitter, such as the types of tools, cutting tools and inspecting tools. Besides, you will learn the safety knowledge. Moreover, you can do the process analysis of some simple parts. 可以学习到手动加工的知识和技能，例如，钳工常用工具、刀具、量具的种类及使用方法；操作中的安全常识。另外，能进行一些较简单零件的工艺分析。

④ It's called bench drilling machine. The operation skills of bench drilling machine is essential for CNC Technology majors. 这是台式钻床。台式钻床的操作对数控技术专业的学生是必需的。

Activity 4: Group work. What can you learn in Manual Machining Workshop? 小组活动。你在手动加工车间学习哪些内容？

1. tools of _____
2. safety _____
3. skills of _____
4. process _____
5. operation of _____

Activity 5: Oral practice 口语训练

1. Role play this dialogue with your partner.

2. Discussion: Do you think it is necessary to set a classroom in the workshop?

Activity 6: Translate the English words into Chinese according to the following picture.
根据下图将下列英语单词翻译成汉语。

Hand Tools in a Workshop

1. adjustable spanner _____   2. pliers _____
3. hammer _____   4. Allen Key _____
5. double open ended spanner _____   6. hand saw _____
7. screwdriver _____   8. file _____

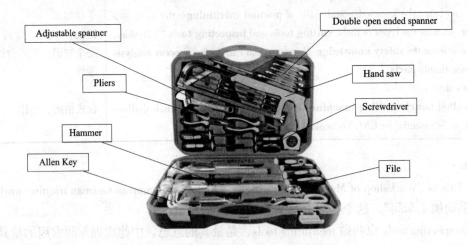

Dialogue 2 Workshop of Machine-tool Processing 机床加工车间

C: We've got **lathe**, **milling machines** and **grinding machines**① here in this workshop. You should learn **professional skills** here, like the **machine operation**, safety requirements and maintenance of machines.	车床，铣床，磨床 专业技能，机床操作
W: How long will we spend to master so much knowledge?	
C: At least 5 weeks theory-learning and **continuous practical training**.	连续实训
W: The operation of the milling machine is the first year's training task, isn't it?②	
C: Yes, it is. You will accomplish it when you are freshmen. The **skill training** should match with the theory learning.③	技能训练
W: Mr. Chen, what are the machines over there?	
C: They are **CNC lathe**, **CNC milling machine** and **machining centers**④.	数控车床，数控铣床，加工中心
W: What will we learn with them?	
C: The engineering **application software**, **to be exact**, Auto CAD, UG etc.	应用软件，准确地说
W: **Terrific**!	太棒了

Note:
① lathe 是车床，也可以叫做 turning machine，是最早的机床之一。milling machine 是铣床，是用途最广泛的机床。grinding machine 是磨床。

② The operation of the milling machine is the first year's training task , isn't it? 本句中包含一个反义疑问句。反义疑问句加在陈述句之后，对陈述句表达的事实或观点提出疑问。陈述句

部分如果是肯定句，反义疑问句的动词部分要用否定形式。

③ The skill training should match with the theory learning. 这些技能培养还要配合理论的学习。

④ CNC lathe 是数控车床，CNC milling machine 是数控铣床，machining center 是加工中心。CNC 是首字母缩略词，是 Computerized Numerical Control 的缩写。

Activity 7: Choose sentences from ①–⑤, and finish the dialogue. 将句子①~⑤填入对话中合适的位置，完成对话。

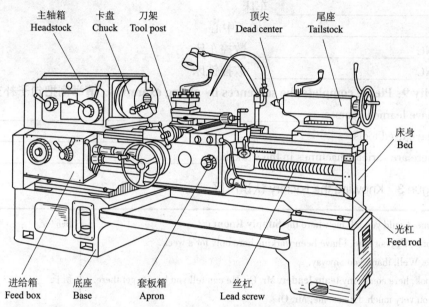

Mr.Li: 1. _____. Do you know the structure of the lathe?

Bob: 2. _____

Mr.Li: What's this?

Bob: It's a headstock. Sometimes people call it as the heart of a lathe.

Mr.Li: 3. _____

Bob: It's a chuck and the work piece can be mounted on it.

Mr.Li: What's that?

Bob: It's tool post and carriage. The cutting tools are put here.

Mr.Li: What's that?

Bob: It's center and tailstock.

Mr.Li: What's that?

Bob: It's a bed. These are all the main structure of the lathe. 4. _____

Mr.Li: Yes. You are very kind.

Bob: 5. _____

① Excuse me.

② Are those clear to you?

③ Yes. I know some of them.

④ What's this then?
⑤ Delighted to have been of assistance.

Activity 8: Complete the following words according to the Chinese meanings. 根据汉语意思将单词补充完整。

1. m _____ 铣床
2. g _____ 磨床
3. d _____ 钻床
4. l _____ 车床
5. m _____ 加工中心
6. CNC _____ 数控车床
7. CNC _____ 数控铣床

Activity 9: Please complete the sentences by giving examples. 请举例把句子补充完整。

1. I have learned courses _____.
2. I have used tools _____.
3. There are various machines in the workshop _____.

Dialogue 3　Knowing the Factory 认识厂区

Z: Excuse me. Could you tell me where the **Supply Room** is?	供应室
Worker: Sorry, I'm not sure. I have been working here only for a week.	
Z: Oh, I see. Well, thank you anyway.	
Worker: Look, here comes my **team leader**, Mr. Qi. He can tell you how to get there.	班长
Z: Thank you very much. Excuse me, Mr. Qi.	
Q: Yes, I don't think I have seen you before. Are you new here?	
Z: Yes, this is the second day I have been working here. My name is Zhang Yang, and I work in the **Assembly Shop**.	装配车间
Q: Oh, I see. What can I do to help?	
Z: I'm looking for the Supply Room. Could you tell me where it is?	
Q: Certainly. Let me see... ①Go straight on until you come to the **Training Center**. Then turn right, and you will see the Workshop②. The supply room **is opposite to** it and near the **Maintenance Division**③.	培训中心 对面 维修部
Z: I think I will find it. Thank you for your help.	
Q: You're welcome.	

Note：
① Let me see... 常用于句与句之间，可以翻译成"让我想想"。
② 指路相关句型：
Walk along this road/street. 沿着这条路 / 街走。
It's about ... meters from here. 从这里大约……米。
Take the first/... turning on the left/right. 在第……个转弯处左 / 右转。
It's about ... meters along on the right/left. 沿右边 / 左边大约……米。
Walk on and turn left/right. 继续走再向左 / 右转。

Turn right/left at the traffic lights. You'll find the ... on the right/left. 在交通灯右 / 左转，你会发现……在右 / 左边。

Go on until you reach the end of the road/street. You'll see the ... in front of you. 继续走一直到路 / 街的尽头，你就会看到……在你的面前。

Go down this street/road until you reach the 1st/2nd/... traffic lights. Turn right/left. At the end of the road/street you'll see the ... 沿着这条路向前走到第一 / 二个交通灯。右 / 左转。你会看见……在道路的尽头。

③ 方位相关表达：

There are...main parts in our factory. They are...

...is located in the middle of our plant.

...is behind...

...is next to the...

...is in front of the...

...is near the...

...is on the opposite side of...

Activity 10: Complete the following passage according to the dialogue. 根据对话完成下面的短文。

Zhang Yang is a signed _____ and works in _____. On the _____ day of his internship, he wanted to go to the _____ but lost his way in the factory. He asked a _____ for help. Unfortunately, the worker couldn't help him because _____. Finally, a _____—Mr. Qi told Zhang Yang the way and the supply room is opposite to the _____ and near the _____.

Activity 11: Oral practice 口语训练

1. Describe the route from our campus to your signed corporation.
2. Do you think the e-map is reliable? Why or why not?

❋ Reading and Writing 读写拓展

Reading 阅读

Wilnsdorf is a municipality in the district of Siegen–Wittgenstein, in North Rhine–Westphalia, Germany. The southern municipal limits, with the Kalteiche peak, part of the Rothaargebirge, form not only the community's highest point, at 579 m above sea level, but also the boundary between Hesse (Lahn–Dill–Kreis) and North Rhine–Westphalia. The cultural history meeting place affords one a "trip through time": from geology through the Stone Age and the ancient civilizations and into the Middle Ages and modern times.

SIEGENIA GROUP offers a wide range of products which add value to windows and doors. With its TITAN, ALU and PORTAL product groups, together with the AERO and DRIVE ones, the company is one of the world's leading suppliers of hardware, ventilation and building technology. In addition to that, KFV provides a complete range of modern locking systems for doors.

hardware 硬件，五金

ventilation technology 通风技术

Activity 12: Read the following passage and finish the questions below. 读短文回答问题。

What is GPS?

The Global Positioning System (GPS) is a location system based on a constellation of about 24 satellites about 11,000 miles above. GPS was developed by the United States Department of Defense (DOD), for its tremendous application as a military locating utility. The DOD's investment in GPS is immense. Billions and billions of dollars have been invested in creating this technology for military uses. However, over the past several years, GPS has proven to be a useful tool in non-military mapping applications as well.

GPS satellites are orbited high enough to avoid the problems associated with land based systems, yet can provide accurate positioning 24 hours a day, anywhere in the world. Uncorrected positions determined from GPS satellite signals produce accuracy in the range of 50 to 100 meters. When using a technique called differential correction, users can get positions accurate to within 5 meters or less.

Today, many industries are leveraging off the DOD's massive undertaking. As GPS units are becoming smaller and less expensive, there are an expanding number of applications for GPS. In transportation applications, GPS assists pilots and drivers in pinpointing their locations and avoiding collisions. Farmers can use GPS to guide equipment and control accurate distribution of fertilizers and other chemicals. Recreationally, GPS is used for providing accurate locations and as a navigation tool for hikers, hunters and boaters.

Many would argue that GPS has found its greatest utility in the field of Geographic Information Systems (GIS). With some consideration for error, GPS can provide any point on earth with a unique address (its precise location). A GIS is basically a descriptive database of the earth (or a specific part of the earth). GPS tells you that you are at point X,Y,Z while GIS tells you that X,Y,Z is an oak tree, or a spot in a stream with a pH level of 5.4. GPS tells us the "where". GIS tells us the "what". GPS/GIS is reshaping the way we locate, organize, analyze and map our resources.

1. GPS refers to _____.
2. GIS refers to _____.
3. GPS was originally developed for the use of _____.
   A. military   B. agriculture
   C. transportation   D. recreation
4. According to the passage, why is GPS used more than before?
   A. Because farmers make use of GPS widely.
   B. Because hikers make use of GPS widely.
   C. Because drivers can use GPS to find directions.
   D. Because GPS becomes smaller and cheaper.
5. What's the accuracy of positions provided by GPS satellites signals?
   A. 5-10 meters.   B. 50-100 meters.

C. 24 hours.   D. less than 5 meters.

Writing 写作

Training Application Form 培训申请单

填单日期: Date:	年 月 日				
申请人 Applicant		部门 Department		职务 Title	
受训日期 Period	年 月 日至 年 月 日 From to				
课程名称 Training Course			讲师 Instructor		
举办单位 Organizer/Institution			受训地点 Location		
受训人员 Participants					
培训费用 Training Fee					
训练目的: Aims of Training					
主管审核: Supervisor Review			总经理批准: Approval by GM		
签名: Signature	日期: Date		签名: Signature	日期: Date	

Application Letter 申请信

Application letter includes cover letter, application letter for studying abroad, etc. Writing an Application Letter should pay attention to the short and concise language, sincere and courteous tone, and be sure to avoid exaggeration.

申请信类型很多，包括求职信、留学申请信等。写申请信应注意言简意赅，语气诚恳礼貌，避免夸张。

The application letter generally includes: the reason for the application, the conditions, and express their gratitude and hope that the other side should consider and reply the application.

申请信的内容一般包括：申请原因、具备条件、表示谢意，恳请申请单位考虑自己的申请，期盼回复。

The first paragraph introduces the incident and explains the purpose of the letter. The second paragraph is a detailed introduction to the project applied. The third paragraph makes a request and expresses gratitude for the reply.

第一段引入事件，说明写信目的。第二段介绍所申请项目的具体情况。第三段提出请求，表示感谢盼复。

1. 写申请信的目的常用句型

I am writing to you in the hope that I may obtain an opportunity to ....

I would like to apply ...

2. 介绍个人优势常用语

I am good at English and especially my spoken English is fairly good.

I am currently studying ...

3. 表示感谢及请求常用语

Thank you for considering my application and I am looking forward to your early reply.

I greatly appreciate any of your favorable consideration of my application.

I would be very grateful if you could consider my application.

Activity 13: Please finish a Letter of Application according to the context. 请根据情境完成申请信。

王楠想申请质检部秘书职位，因为她对质检工作很感兴趣并且有英语方面的优势。请帮助她完成申请信。

Dear Sirs,

　　I am extremely pleased to notice the ＿＿＿＿＿＿＿＿＿＿＿＿＿＿＿＿＿＿＿（报纸上的广告） for the ＿＿＿＿＿＿（招聘）. And I'm writing to apply for the＿＿＿＿＿＿＿＿（质检部秘书）.

　　I am confident that I am suitable for the＿＿＿＿（职位）. On one hand, ＿＿＿＿＿＿＿＿＿＿（原因）.On the other hand, ＿＿＿＿＿＿＿＿＿＿＿＿＿＿＿（原因）.

　　I shall be much honored if you will offer me the opportunity to＿＿＿＿＿＿＿（尝试）. I am looking forward to your reply at your earliest convenience.

　　　　　　　　　　　　　　　　　　　　　　　　　　　　　　Yours Sincerely,

　　　　　　　　　　　　　　　　　　　　　　　　　　　　　　Wang Nan

Activity 14: Please finish a Letter of Application according to the context. 请根据情境完成申请信。

Context：质检部想组织一次全体员工培训，质检部办公室主任 Mr. Johnson 写了一份培训申请提交给总经理 Evan Thomas。培训内容是精益管理，时间为3月20、21日两天，费用共计8 000元。

_____
_____
_____
_____

 Speaking Skills 交际技巧

话语填充

很多同学在写作文时，总会有源源不断的灵感，而在口语陈述时，却不知道如何开口表达，断断续续词不达意。有时候会没有思路，不知道如何表述，有时候话讲到一半就突然卡住了。在日常交际的时候，为了不影响流利度，同时还不能出现冷场，要给自己的思考留出喘息之机，我们往往需要一些话语来过渡，话语的长短视你需要思考的时间而定。掌握一些比较好的词或者句子，往往能够帮助我们避免尴尬。

1. 表达不出时的对话填充物

1) How should I put it? 我该怎么说呢？

This phrase shows that you are looking for the best expression. It also gives a heads-up that you are about to say something that might be a bit tricky to explain. 这句话让人感觉你是在寻找最恰当的表达方式，同时也提醒对方你接下来要表达的内容是复杂的。

2) What's the word I am looking for? 什么词最能表达我的意思呢？

拓展：What's the phrase/expression I am looking for? 这句话表明了你是真的在认真思考，说这句话通常不是真的在提问，而是给自己时间过渡。

3) It's on the tip of my tongue. 我想说的话就在嘴边。

"The tip of my tongue" 特别形象，好像词真的就在你的舌尖，马上要蹦出来一样。

4) I just had it. 我刚才还记得的。

不恰当举例：I forgot.

5) It's just not coming to me. 这个词就是怎么都想不起来啊。

拓展：It slipped my mind.（我疏忽了/忘了。）

2. 简单过渡词句

如果你不需要很长的思考时间，可以说些简单的过渡词：

| "well" |
| "you know" |
| "actually" |
| "I mean" |
| "personally" |
| "to be honest" |

"on the other hand"

"frankly"

"as a matter of fact" 等,

这些表达也叫"discourse marker"(语篇标记),也就是并不改变句子本质意思的语言填充物。这样的表达可以帮助我们争取思考的时间。

这些表达因人而异,还可以选择如下:

1) Well, the first thing that comes to my mind is that...

2) Okay, there are several reasons to consider...

3) I am not an expert in this filed, but as far as I know, ...

3. 过渡句

如果问题有难度,需要更多的思考时间,那么可以先针对问题给出一个简短的评价,为自己争取更多的时间:

1) That's difficult to answer, but (maybe)...

这个问题很难回答,但是(也许)...

2) I'm sorry, but I don't know much about...but perhaps...

不好意思,但是我对于……不是很了解,但是也许……

3) Maybe I can answer your question by telling you about a personal experience I had. 也许我可以通过告诉你一个我自己的经历来回答你的问题。

4) That's an interesting question...Let me see. Well, I suppose that...

这是一个很有趣的问题……让我想想,嗯,我认为……

5) Yes, that's a big issue. 是的,那是一个大问题。

Activity 15: Please underline the expressions functioned as fillers. 请划线下列句子中起填充作用的词或句子。

1. Well, you know. It is the computer that has made it possible.

2. Okay, it's a clever question. Here you mix a robot with an ordinary washing machine.

3. How should I put it? Let me see... Well, when we talk about industrial robots, we mean a machine or device, which can perform a factory duty in a stand-alone manner.

4. I'm not an expert in this field, but as far as I know, some reports are inconsistent with facts.

5. And to be honest, we did have a good time playing a board game until his friends came into the room.

Self-evaluation

Items	Yes	No
I have acquired all the key words in the Mission Objective		
I have acquired all the sentence patterns in the Mission Objective		
I can compose an Application Letter		

Unit 5

Health and Safety 健康与安全

✽ Warming-up 课程准备

Mission Objective 任务目标

1. 掌握安全、危险、禁止、防护、安全标识、安全意识、规章制度等安全用词和灭火器、护目镜、安全鞋、安全帽、手套、耳塞等常用安全设施，火灾、触电、摔倒、跌落等安全隐患的英语表达。

2. 会用英语复述灭火器的使用方法、询问和回答常见安全标识的含义、谈论办公室情境中与安全有关的话题。

3. 会填不合格品报告单、会撰写英文电子邮件。

Discussion 讨论

1. What safety regulations do you know?
2. Do you know any safety equipment? Why do you need them?

Words & phrases for assistance:

安全: safety, security, dangerous, safety signs, safety awareness, regulations, forbidden, not allowed, hazards, risk, safety equipment, inspections

Database 知识库

5 Common Office Hazards and How to Avoid Them

1. Slips and Falls

Slips and falls can be caused by something as simple as a wet floor. Other causes of slips and falls can include uneven floors and work surfaces and cluttered walkways and work spaces. These can be countered by ensuring wet floor signs when needed and organizing shared corridors and places of high traffic.

2. Fire Hazards

Fire safety is extremely important in every workplace, as you don't want to see all your work go up in flames! Fortunately, it can be avoided by taking the smallest of precautions. Check all your power cords and replace any that are fraying, avoid overloading power boards and outlets, and monitor the use of space heaters closely. Additionally, ensure all employees know where fire extinguishers are and make sure emergency exits are clear at all times.

3. Eyestrain

Computers are now used in the majority of workplaces for almost everything, so it is not surprising that eyestrain is becoming increasingly common in office jobs. Eyestrain may cause eyes to become irritated and dry, which will ultimately lead to a loss of concentration. Taking regular breaks from the screen and ensuring the lighting is appropriate for the task at hand are a couple of ways to prevent eyestrain.

4. Stress

Stress is a more general office hazard, but it can have quite a significant impact on both your work and home life. Everyone has different ways to reduce their stress levels, but common strategies include taking consistent breaks, avoiding working overtime and being organized in your work.

5. Ergonomics

Incorrect use of office furniture can lead to a variety of long-term physical injuries. These can affect everything from your wrists to your neck and back. Be sure to set up your workstation to suit your proportions.

Activity 1: Fill in the blanks according to the passage. 根据短文填空。

1. The five common hazards in the office are ____, ____, ____, ____, and ____.
2. Making sure all employees know where fire extinguishers are can avoid _____.
3. Stress can have quite a significant impact on both your _____ and _____.

Activity 2: Please complete the table according to the passage. 根据短文完成表格。

Hazards	Causes	Preventative Measures
Slips and Falls		
Fire Hazards		
Eyestrain		
Stress		
Ergonomics		

🎯 Oral Practice 口语训练

人物涉及 (Characters):

Q: Mr. Qi (a Team Leader of Logistics Department)　物流部班组长

W: Wang Nan (an Apprentice from Jinan Vocational College)　济南职业学院学徒

R, A: Rose, Amy (Colleagues in the Same Company)　同事

H: Henry (a Senior Staff)　老员工

情景解说 (Situations):

学生到企业参观，公司物流部班组长 Mr. Qi (Q) 向学生讲解该公司的防火措施和灭火器的使用方法。

Rose(R) 在工作中意外摔伤了手腕，培训中心为此进行了安全培训。

Wang Nan(W) 在操作机床时，Henry(H) 发现了安全隐患，及时提醒她。

以下三个对话围绕三个情境展开。

Dialogue 1　Using Fire Extinguishers 使用灭火器

Q: Have you seen the red bottle there?	
W: Yes, **fire extinguisher,** the **equipment** to fight the fire.	灭火器，设备
Q: You can see them at every corner of the **shop**. It is easy for you to remember① how to use the fire extinguisher if you remember PASS.	车间
W: That's very interesting. Could you explain?	
Q: PASS stands for② **pull, aim, squeeze** and **sweep**③. To be exact, pull the **pin** first. Then, **aim at**④ **the base of fire**. Third, **squeeze the top handle**. The last step is to **sweep from side to side**.	拔，对准，挤压，喷扫，拴 对准火的底部 挤压顶部把手 左右喷扫
W: The picture is **vivid**.	生动的、鲜明的

- Pull
- Aim
- Squeeze
- Sweep

the pin

the handle

side to side

Q: Although we know how to operate the fire extinguishers, **in case of** a fire alarm, we still need to call 119 and escape to the **assembly point** as soon as possible.	万一 集合点
W: That sounds **reasonable** and I will follow your instruction.	合理的

Note：

① It is easy for sb. to do sth. 对某人来说做某事很容易。固定搭配 It is + adj. + for + sb. (to do) 表示对某人来说做某事太……例如：It's impossible for me to leave my family. 我是不可能离开家的。也可以变形为 It is + too + adj. + for +sb. (to do) 表示否定。例如：It's too expensive for me to buy. 太贵了，我买不起。

② stand for 这个短语的意思是"表示、意味；代表、象征"；例如：The sign of △ in the biology stands for the evergreen plant. 在植物学上△符号表示四季常青植物。The olive branch stands for peace. 橄榄枝象征着和平。

③ 在文中 PASS 是一个首字母缩略词（acronym），即通过组合每个词的首字母构成新词或专有名词。例如，OPEC 是石油输出国组织 (Organization of Petroleum Exporting Countries) 的首字母缩略词。

④ aim at doing sth./ aim at sth. 是"瞄准，针对"的意思。例如，He is aiming at the target. 他在瞄靶。Businesses will have to aim at long-term growth, not the present profit. 商业应立志于长远发展，而不是眼前利益。常用的还有 aim to do sth. 表示想要做某事，立志要做某事。

Activity 3: Can you write down the full names of the following acronyms? Look up a dictionary if necessary. 你能写出下列首字母缩略词的全称吗？可以借助词典完成。

1. VIP 贵宾 ＿＿＿＿＿＿＿＿＿＿＿＿＿＿＿＿＿＿
2. UN 联合国 ＿＿＿＿＿＿＿＿＿＿＿＿＿＿＿＿
3. MVP 最有价值球员 ＿＿＿＿＿＿＿＿＿＿
4. IT 信息技术 ＿＿＿＿＿＿＿＿＿＿＿＿＿＿
5. CPU 中央处理器 ＿＿＿＿＿＿＿＿＿＿＿
6. CAD 计算机辅助设计 ＿＿＿＿＿＿＿＿
7. PLC 可编程逻辑控制器 ＿＿＿＿＿＿＿
8. CNC 计算机数字控制 ＿＿＿＿＿＿＿＿
9. DC 直流电 ＿＿＿＿＿＿＿＿＿＿＿＿＿＿
10. AC 交流电 ＿＿＿＿＿＿＿＿＿＿＿＿＿＿

Activity 4: Rearrange the words and make them a meaningful sentence. 重新排列词语构成完整的句子。

1. useful English It to is learn for well us
＿＿＿＿＿＿＿＿＿＿＿＿＿＿＿＿＿＿＿＿＿＿＿＿＿＿＿＿＿＿＿＿＿＿＿＿

2. him is It know important for to this
＿＿＿＿＿＿＿＿＿＿＿＿＿＿＿＿＿＿＿＿＿＿＿＿＿＿＿＿＿＿＿＿＿＿＿＿

3. impossible finish It for is us to the work
＿＿＿＿＿＿＿＿＿＿＿＿＿＿＿＿＿＿＿＿＿＿＿＿＿＿＿＿＿＿＿＿＿＿＿＿

4. It for us to to on get school is time necessary
＿＿＿＿＿＿＿＿＿＿＿＿＿＿＿＿＿＿＿＿＿＿＿＿＿＿＿＿＿＿＿＿＿＿＿＿

5. It the him to exam is hard for pass
＿＿＿＿＿＿＿＿＿＿＿＿＿＿＿＿＿＿＿＿＿＿＿＿＿＿＿＿＿＿＿＿＿＿＿＿

Activity 5: Oral practice 口语训练

1. Role play this dialogue with your partner.
2. Can you find fire extinguishers in your workshop? Can you describe how to use them?

Dialogue 2 An Accident in the Office 办公室事故

R: Amy, are you coming for lunch?	
A: Yes, but I have to **put** these files **away**① first.	收好
R: I heard that you got an accident last week. How are you now?	
A: I **fell off** the chair and hurt my **wrist**. Luckily, it's nothing **serious**.	跌落，手腕，严重的
R: Have you got **first aid**?	急救措施
A: Yes, Jimmy gave me first aid **immediately**. I **took** the week **off**②, and went home.	立即，请假
R: It's because of the accident that the training center gave a safety **lecture** yesterday.③	讲座
A: It's a pity that I **missed** it. Can you give me some **hints**?	错过，线索
R: He has **mentioned** common hazards in office area and **measures** to prevent them, **especially** fire hazards, **electric shock**, etc. It is too **complicated** to be explained in a few words.④	提及，措施 尤其，触电，复杂的
A: Please go into detail after lunch.⑤	
R: Sure.	

Note：

① put away 放好；抛弃；储存。

put 的相关词组：

put out 熄灭；伸出；出版；打扰；put up 提供；建造；举起；推举；put off 推迟；扔掉；阻止。

② take off 在本句中是"休假，休息"的意思。例如，Both students and teachers took Christmas off. 学生和老师圣诞节都休假。另外，take off 也表示出发、匆匆离开。例如，The professor took off for Shanghai this morning. 教授今天上午动身去上海了。

③ It's because of the accident that the training center gave a safety lecture yesterday. 这是一个强调句，强调原因状语。"正是由于事故，培训中心的工作人员昨天给我们进行了安全知识讲座。"强调句的基本结构是 It is/was + 强调部分 +that+ 其他成分。构成强调句的 it 本身没有词义；强调句中的连接词一般只用 that 和 who，即使在强调时间状语和地点状语时也如此，并且 that 和 who 不可省略；强调句中的时态只用两种，一般现在时和一般过去时。例如，针对 I met Li Ming at the railway station yesterday. 句子成分进行强调。

强调主语：It was I that (who) met Li Ming at the railway station yesterday.

强调宾语：It was Li Ming that I met at the railway station yesterday.

强调地点状语：It was at the railway station that I met Li Ming yesterday.

强调时间状语：It was yesterday that I met Li Ming at the railway station.

④ It is too complicated to be explained in a few words. 这句话是说"这很复杂，三言两语说不清楚。"

⑤ Please go into detail after lunch. 这句意思"午饭后请给我详细解释一下"。go into detail 是详细解释的意思。

Activity 6: Please choose the best answers. 选择。

1. It is the ability to do the job _____ matters , not where you come from or what you are.

A. one   B. that   C. what   D. it

2. It is a winter night _____ he spent with me.
A. that   B. where   C. as   D. when
3. It was without saying goodbye _____ her father left her.
A. that   B. which   C. when   D. in which
4. It was _____ it was raining so hard that we had to stay at home all day.
A. since   B. for   C. as   D. because
5. It was what he said _____ disappointed me.
A. that   B. what   C. when   D. where

Activity 7: Filling in the blanks using the words in the table. 用下表中的词或短语填空。

too...to...	for instance	take...off	put...away

1. He is _____ slow _____ meet the schedule.
2. He _____ two weeks _____ in August.
3. Please _____ your dishes _____ after dinner.
4. There are jobs more dangerous than truck driving, _____, fire fighting.

Activity 8: Oral Practice. 口语训练。

1. Do you know the hazards around you in your workshop?
2. Do you know how to prevent the hazard?

Dialogue 3　Safety Instructions in Workshop　车间安全指导

Wang Nan is working with a lathe and Henry comes into the workshop.

H: Good morning, Wang Nan.	
W: Good morning, sir.	
H: Wang Nan, I have something to **remind** you.	提醒
W: I am all ears.①	
H: What does the **sign** on the wall mean?	安全标识
W: It means **safety glasses**② must be worn.③	护目镜
H: Not **exactly**. It means safety glasses must be worn all the time.	精确
W: I am sorry. I got it. I will follow the sign **strictly**.	严格地
H: Safety is the most important thing for us. We should be more careful, right?	
W: Yes, I see.	
H: I hope you work carefully and **responsibly**. After all, **neither** of us ever wants an accident.	负责任地，两者都不
W: I will keep it in mind④. Thank you very much.	
H: That's all right. Hope you enjoy your work.	

Note：

① I am all ears. 是一个习惯用法，意思是"洗耳恭听"。

② safety glasses，也叫做 safety goggles，是机械加工车间常用的安全护具。此外，常用的安全防护用品还有安全鞋（safety boots）、安全帽（hard hats）、手套（safety gloves）、工作服（uniform），有的岗位还需要耳塞（ear plugs）防止噪声损害听力。

③ "–What does the sign mean?–It means..." 是询问某个标识的含义的句型。

④ keep it in mind 也可以说 bear it in mind，是"记住"的意思，比 I'll remember it 更口语化。

Activity 9: Match the safety signs with their meanings. 将安全标识与含义连线。

A. Exit	B. Fire Hose	C. Fire Extinguisher
D. Do not use mobile phones.	E. Fire Alarm Call Point	F. Assembly point
G. Safety gloves must be worn.	H. No Drinking.	I. High Voltage
J. No Smoking.	K. Uniform must be worn.	L. Forbidden to wear high heels.
M. No Touching.	N. Hard hats must be worn.	O. Earplugs must be worn.
P. Safety boots must be worn.	Q. No photograph.	R. Safety goggles must be worn.

Activity 10: Oral practice. 口语训练。

Have you seen safety signs in your workshop? Practise describing their meanings using the following sentence patterns.

A: What does the sign mean?
B: It means _____.
A: I got it./ I see. I'll follow the sign strictly.

Activity 11: Classify the 4 safety signs according to the information in the table. 学习下表中安全标识的知识，给下列安全标识分类。

Contents	Prohibitory Sign	Warning Sign	Information Sign	Direction Sign
Meaning	Stop	Caution	Safety	Mandatory
Color	Red	Yellow	Green	Blue
Contrast color	White	Black	White	White
Color of graphic symbol	Black	Black	White	White
Shapes	Circle	Triangle	Square	Circle
Application	1 ( )	2 ( )	3 ( )	4 ( )

A

B

C

D

Reading and Writing 读写拓展

Reading 阅读

Nuremberg is a city in the German state of Bavaria, in the administrative region of Middle Franconia, about 170 kilometres (110 mi) north of Munich. It is the second-largest city in Bavaria (after Munich). As of February 2015 it had a population of 517,498, making it Germany's fourteenth-largest city.

DIEHL Metering (DIEHL Metering (Jinan) Co., Ltd.) is the leading heat meter, water meter, gas meter and automatic meter reading system supplier.

  heat meter 热量表
  water meter 水表
  gas meter 燃气表
  automatic meter reading system 远程抄表系统
  supplier 供应商

Activity 12: Read the following passage and finish the questions below. 读短文回答问题。

Garden work could mean trouble if you don't take proper precautions. When using any garden tool, we have these safety tips:

Dress appropriately for the work environment; Wear long pants and long-sleeved shirts to provide some protection from thrown objects; wear close-fitting clothes and don't wear anything that could get caught in moving parts, e.g. loose jewelry; wear sturdy shoes with slip resistant rubber soles.

Walk around the area in which you will be working before starting lawn and garden work, and remove any objects that could damage equipment or cause injury or property damage.

Keep children indoors away from power equipment. Children move quickly and are attracted to mowing and other power equipment activity.

Be sure that safety devices on the equipment are in place and functioning properly before starting work.

Unplug electric tools and disconnect sparkplug wires on gasoline-powered tools before making adjustments or cleaning jams near moving parts.

Be sure power tools are turned off and made inoperable if they must be left unattended. This will prevent use by children.

Never smoke around gasoline.

Never let young children operate power lawn and garden equipment. Teenagers should only be allowed to operate outdoor power equipment if they possess adequate strength and maturity to do so safely. They also should be supervised by a responsible adult.

Never work with electric power tools in wet conditions.

1. What is NOT appropriate to wear in the work environment?

    A. Long Pants.   B. Long-sleeved shirts.

    C. Close-fitting clothes.   D. Loose jewelry.

2. We should _____ before starting lawn and garden work somewhere.

    A. walk around the area

    B. remove any objects that could damage equipment

    C. wear hearing protection

    D. both A and B

3. What should we do with power tools if they must be left unattended according to the passage?

    A. Be sure power tools are turned off.

    B. Keep children indoors.

    C. Never smoke.

    D. Never let young children operate them.

4. _____ is prohibited near gasoline.

    A. No smoking   B. Smoking

    C. Working with electric power tools   D. A wet extension cord

5. We can never work with electric power tools in _____ conditions.

    A. wet   B. any   C. hot   D. cold

Writing 写作

Non-conformance Report 不合格品报告

Source 来源	☐ Purchased 采购件 ☐ In Process 过程中 ☐ Stock 库存			Order/Batch No.: 订单号 / 批号	
	Part No. 零件号		Part Description 零件名称	Supplier 供方	
Lot Size 批量	Sample Size 样本大小		Non-conformance Qty. 不合格数量	Responsible Party 责任方	
	Specification 技术要求	Inspection Result (Defects) 检查结果(缺陷)		Qty. 数量	Sketch 草图
1					
2					
Inspector 检验员			Date 检查日期		

根本原因 Root cause	Technical flaws 技术原因 ☐ YES ☐ NO Repetition 是否重复发生 ☐ YES ☐ NO
	Supervisor/Date 负责人 / 日期:

处理方式 Disposition	1. Corrective action in short time 短期纠正措施:
	NO. of Producing/ 在制品数量 _____, Operator/ Date 执行人 / 日期: _____; NO. of Stock/ 库存数量 _____, Operator/ Date 执行人 / 日期: _____。
	2. Inform the suppliers to take some action 是否需要供应商采取相应行动: ☐ NO ☐ YES, Operator/ Date 执行人 / 日期_____
	NOTE 备注:
	Supervisor/Date 负责人 / 日期:
	Results 结论: ☐ Scrapping 废弃 ☐ Redoing 返工 ☐ Repairing 维修
	NOTE 备注:
	Supervisor/Date 负责人 / 日期:
	Approved by GM/Date 总经理审批 / 日期:

E-mail 电子邮件

The basic elements of an English e-mail are the Subject, Appellation, Body, Ending phrase, and Signature.

英文电子邮件的基本要素是主题、称谓、正文、结尾用语及署名。

1. Subject

The most important part of the e-mail is the subject. The topic should be concise and highlight the importance of the e-mail. The content in the Subject Box should be a concise summary of the content of the letter. The theme can be a word, a phrase or a full sentence, but it should be less than 35 letters. The subject should not be vague.

1. 主题

电子邮件最重要的部分是主题，主题应当做到言简意赅并突出邮件重要性。主题框的内容应简明地概括信的内容，可以是一个单词、短语，也可以是完整句，但长度一般不超过35个字母。主题框的内容切忌含糊不清。

2. Appellation

If it is the first time to write to someone, it is better to call them "Dear+ full name", which may make people feel more formal. If the other party is writing in an informal manner, we can reply informally. For example: "Hello/Hi Lillian".

In actual communication, you may not know the other person's name. You can use "Dear+ title", such as "Dear President", or "Dear Sir/Madam".

2. 称谓

如果是第一次给对方写信，那么称谓最好用"Dear+全名"，这样会让人感觉比较正式。如果对方以非正式口吻来信，我们也可以类似非正式地回复。比如："Hello/Hi Lillian"。

在实际通信中可能遇到不知道对方姓名，可以用"Dear+对方头衔"，如"Dear President"，或者"Dear Sir/Madam"的形式。

3. Body

When writing the body, write the most important thing in the first paragraph when the content of the mail is long. In order to make the recipient more comfortable to read the e-mail, we need to pay attention to the aesthetic of the body structure of the e-mail, which is best controlled within two or three paragraphs. If an e-mail involves multiple information points, we can use item-by-item method, such as symbols, headings and numbers, to make the content of the message to be expressed clearly.

3. 正文

在书写正文时，把最重要的事情写在第一段。为了让收件人阅读邮件比较舒服，我们需要注意邮件正文结构的美感，邮件段落最好控制在两三段之内。如果一封电子邮件涉及多个信息点，我们可以采用分条目的方法，如用符号、小标题、编号来使得邮件想要表达的内容层次清晰。

4. Ending Phrase

The ending phrase is added after the body. Note that only the first word in the ending phrase is capitalized and all remaining words are lowercase. The selection of different ending phrase is based on the relationship between the writer and the recipient.

Very Formal: Respectfully yours, Yours respectfully

Formal: Very truly yours, Yours very truly, Yours truly

Less Formal: Sincerely yours, Yours sincerely, Sincerely, Cordially yours, Yours cordially, Cordially

Informal: Regards, Warm regards, With kindest regards, With my best regards, My best, Give my best to Mary, Fondly, Thanks, See you next week!

4. 结尾

结尾语在正文之后添加。注意一般结尾语中只有第一个单词首字母大写而剩余单词都小写。书信的结尾致意根据写信人和收信人的关系选择不同的用语。

非常正式的：Respectfully yours, Yours respectfully

正式的：Very truly yours, Yours very truly, Yours truly

不太正式的：Sincerely yours, Yours sincerely, Sincerely, Cordially yours, Yours cordially, Cordially

非正式的：Regards, Warm regards, With kindest regards, With my best regards, My best, Give my best to Mary, Fondly, Thanks, See you next week!

5. Signature

At the end of the body you need to sign your name, you can write the full name, or only write the name. When you need to identify the gender, you can specify (Mr./Ms.) after the name. For Chinese people, to distinguish the last name and the first name, we can capitalize all the letters of our family name. For example, XIONG Lillian. If the writer represents an organization or department, write down the position and the department of the company under the name.

6. 签名

在正文最后需要署名，可以写全名，也可以只写名字。需要辨明性别时可以在姓名后面注明 (Mr./Ms.)。对于咱们中国人，为了区分姓和名，可以把我们的姓的字母全部大写，例如 XIONG Lillian。如果写信人代表的是一个组织或部门，应在名字下一行写上自己的职位、所属部门。

Samples:

邮件举例：

Job seeking: E-mails for job seeking are roughly divided into several sections: applying for a position, introducing yourself, expecting a reply, and expressing gratitude.

工作求职：工作求职类的邮件大体分为申请职位、介绍自己、期待回复、表达谢意几个部分。

Dear Sir/Madam,

I am writing to apply for the position of ...（职位）as posted in your website.

Introduce yourself...（自我介绍）

Please find attached a copy of my resume for your review.（我的简历请见附件，作为参考。）

I am looking forward to your reply. Thank you very much for your time and consideration.

Yours sincerely,
Lillian

Express gratitude: In daily life, e-mails are frequently employed to express gratitude. In foreign countries, a Thanks Letter will be mailed after participating in a friend's parties or accepting others' gifts. You need to express your sincere gratitude to the recipient.

表达谢意：日常生活中经常会使用到表达谢意的邮件。在国外，参加完朋友邀请去的一次 party，或者接受了别人馈赠的礼物等情况下都需要写一封 Thanks Letter。此类只要能够表达自己对收信人的真挚的感激之情即可。

Dear Alice,
　　Thanks so much for the lovely dinner last night. It was so thoughtful of you. I would like to invite both of you to my house when you are available.

Best regards,
Lillian

Make appointments: In western countries it is common to make appointments before visiting supervisors, customers, professionals.

预约：在西方国家，通常在拜访主管、客户、专业人士之前都需要预约。

Dear Mr./Ms,
　　Mr. John Green, our General Manager, will be in Paris from June 2 to 7 and would like to come and see you, say, on June 3 at 2.00 p.m. about the opening of a sample room there. Please let us know if the time is convenient for you. If not, what time you would suggest.

Yours faithfully,
Lillian

Notice:
通知

Dear Margret, Jackson, and Francis,
　　We reviewed our performance last month, and reached agreements at last week's departmental meeting. We discussed everything from new staff positions to upcoming projects and potential challenges we may face. The meeting minutes have now been compiled and proofread. You can find enclosed the full transcript of the meeting minutes.
　　Please feel free to let me know if you would like to discuss any topic contained in the meeting minutes, or if you notice any discrepancies between the meeting minutes and the actual meeting content.
　　Thanks for your attention!

Sincerely,
Christine
Secretary to Senior Director of Engineering Department

Activity 13: Please write an e-mail according to the following context. 请根据情境写一封电子邮件。

Context: Mr. Black will come to visit your company next month and you are expected to arrange the accommodation for him. Please write an e-mail to ask him the flight number, date of arrival and schedule.

Speaking Skills 交际技巧

巧用替换

当我们用英语交流时，由于词汇量的限制，经常出现某个词或某个句子不会表达的状况，这时不妨考虑换一个词，换一个句式，或者甚至是换一种思维方式，亦即"替换"。英语表达中巧用替换，既可以缓解无词可用的尴尬，也可以使表达更丰富，更地道，使语气更委婉。

1. 常用表达替换

英语中有许多高频用词，大部分学生都能够熟练运用，但使用过多，难免让人有千篇一律、表达单调的感觉，在平时学习中，不妨积累一些常用词的近义词，以使表达更丰富。

以一个大家最常用到的词 good 为例，可以根据具体情境的不同替换为 wonderful, nice, great, brilliant, amazing, fantastic, excellent, terrific, incredible, unbelievable, 等等。

再比如 thank you，为避免千篇一律，也可用 I appreciate that very much。

2. 难词替换

当我们找不到合适的词汇来表达意思时，可以考虑换一个更常用、更简单的词来代替，或者用解释的方法将其表述出来。比如对方听不懂 I love you，如果替换为 I want to kiss you, I want to hug you, I want to show my heart to you, I like you, 等等，对方可能就会理解你想表达的意思了，即用另一种表达方式来表述相同的意思。比如：

millionaire 可替换为 one who has a lot of money

devote all his spare time to reading 可替换为 spend all his spare time in reading

run across/ encounter 可替换为 meet

3. 名词替换动词

与动词相比，名词更具有静态性。假如只是简单陈述一个事实或概念，不需要动词那样生动具体，可用动作名词来代替动词。这种名词化结构在科技英语中尤其常用。

seat 替换 sit, take a seat 代替 sit down

analysis 替换 analyze, 如: He gave an analysis of the problem.

Activity 14: Please find at least five words to replace the following words. 请找出至少5个词来替换下列单词。

1. clever
2. big
3. beautiful
4. happy

Activity 15: Please replace the underlined words in the sentences with expressions in the box. 请用框内的词代替下列句中划线的词,必要时改变形式。

| be supposed to | hit the sack | due to | round the corner | preserve |

1. We must _____ (protect) and save our planet for our children.
2. Jack was so tired that he _____ (sleep) early last night.
3. He arrived late _____ (because of) the storm.
4. He _____ (should) have driven more slowly.
5. The summer vacation is _____ .(coming)

Self-evaluation

Items	Yes	No
I have acquired all the key words in the Mission Objective		
I have acquired all the sentence patterns in the Mission Objective		
I can compose an e-mail		

Unit 6

6S Management 6S 管理

 Warming-up 课程准备

Mission Objective 任务目标

1. 掌握 6S 管理的 6 个英文单词、会解释其含义。
2. 掌握被动语态的结构和 must be 表示肯定的猜测的用法，会用英语客气地询问，能根据具体情境给出 6S 整改建议。
3. 学会合理化建议表的填写方法，学习撰写建议信。

Discussion 讨论

1. Please describe the three pictures and express your opinion.
2. Which enterprise do you want to work in?

Words and phrases for assistance:

人：worker, employee

物品：trash, trashcan, items, tools, table, chair

形容词：dirty, untidy, disorganized, don't care, angry, tidy, clean, nice, bad habits, good habits

动词：throw, kick, leave, obey rules

Database 知识库

An Overview of 6S

The name "6S" comes from the 5 Japanese words that make up the 5 stages of 6S, each one starting with the letter "S". 6S has added the additional "S" to the original 5. Specifically, 6S are Sort, Store, Shine, Safety, Standardize and Sustaining.

The first step of the "6S" process—Sort, refers to the act of throwing away all unwanted, unnecessary, and unrelated materials in the workplace. In this stage we remove unwanted items, leaving only the tools, equipment, components, and machines that are required on a daily basis. Because of sort, simplification of tasks, effective use of space, and careful purchase of items follow.

The second stage of 6S is "Store". We organize all the items in working area—put everything in an assigned place so that it can be accessed quickly, as well as returned in that same place quickly. The correct place, position, or holder for every tool, item, or material must be chosen carefully in relation to how the work will be performed and who will use them. If everyone has quick access to an item or materials, work flow becomes efficient, and the worker becomes productive.

The third stage "Shine", refers to the cleaning of the area to make the area look pretty. Cleaning must be done by everyone in the organization, from operators to managers. You will have a visually more pleasing environment that will serve as a significant marketing tool of promoting your company.

When talking about Safety, we concentrate on safety environment and personnel awareness of safety. That means reviewing every action and each area to ensure that we have not overlooked any potential hazards.

Standardize is the fifth step. This ensures that there are clear standards and everyone uses the most efficient work method. The right tools are in the right place, the correct methods and standards, and a motivated workforce means that you will have a far more efficient and less wasteful working environment.

The 6th stage is usually called Sustaining. This is where we try to make the 6S process part of the company culture to ensure the ongoing implementation and improvement of this initiative. It denotes commitment to maintain or practice the first 5 "S" as a way of life. The emphasis of Sustaining is elimination of bad habits and constant practice of good ones. Once true Sustaining is achieved, personnel voluntarily observe cleanliness and orderliness at all times, without having to be reminded by management.

Benefits of 6S lie in higher efficiency, fewer accidents, higher levels of quality, and fewer breakdowns. What's more, problems within your processes become immediately obvious.

Activity 1: Fill in the blanks according to the passage. 根据短文填空。

1. 6S refers to _____, _____, _____, _____, _____, _____.

2. Sort means _____ the unnecessary items and _____ the daily-used items.

3. Store means _____ all the items in the working area, and store everything in its own position.

4. _____ aims to make the workplace more pretty.

5. _____ means we define standards and regulations to make sure that all the achievements in the previous stages will be maintained.

6. Safety not only stresses less accidents, but also the employees have the _____ _____.

7. _____ is the ultimate goal of 6S management, that is, every employee has the habit of sort, store, shine, safety and standardize.

Activity 2: Which "S" do the following statements belong to. 判断以下的陈述属于 6S 中的哪一项。

Sort, Store, Shine, Safety, Standardize, Sustaining

(　　) 1. Jack put everything in an assigned place so that it can be accessed quickly, as well as returned in that same place quickly.

(　　) 2. Jack threw away all unwanted, unnecessary, and unrelated materials in the workplace.

(　　) 3. Jack cleaned up the workplace and made the area look pretty.

(　　) 4. The trainer made regulations to make sure there are clear standards for doing everything.

(　　) 5. All the trainees constantly practice good habits and avoid bad ones, without being reminded by trainers and supervisors.

(　　) 6. All the trainees concentrate on every action and every area to ensure that they have not overlooked any hazards.

Oral Practice 口语训练

人物涉及 (Characters):

W: Wang Nan, (Freshman Majoring in Industrial Process Automation Technology)　工业过程自动化技术专业新生

T: Tony, (Trainer Works in the Workshop)　车间培训师

情景解说 (Situations):

开学之初，车间培训师 Tony (T) 带领 Wang Nan (W) 等学生了解 6S 管理的含义和在实训场所实施 6S 管理的重要性。

Dialogue 1　Importance of 6S Management 6S 管理的重要性

W: Sir, what is "6S"?	
T: "6S" is a management **philosophy** which is popular in many companies. It was 5S when originally **invented** in Japan in 1955, and the 6th "S"—safety was added to it when it was **introduced** to China.①	理念 发明 引入
W: Oh, it seems that "6S" is an acronym and stands for 6 words that start with the letter "S".	
T: That's right! They are **Sort, Store, Shine, Safety, Standardize** and **Sustaining**.	整理，整顿，清扫，安全，标准化，素养；强调
W: Can you tell me the reason for② **emphasizing** 6S Management before we start to work?	

T: 6S Management aims at making every employee develop the habit of "doing everything", so as to ③ **achieve the purpose** of improving the **overall** quality of work.	达到目标，整体的
W: The quality of the work can be improved by **employing** 6S Management? It's amazing ④! I can't wait to get better understanding of 6S.	采用
T: **In detail** ⑤, 6S helps improve **efficiency**, achieve better quality, **reduce** accidents, **prolong** the service life of the equipment, reduce waste and **promote company image**.	具体说，效率，减少 延长 提升公司形象

Note:

① 本句中有两个被动语态的用法，分别是"safety was added to it"和"it was introduced to China"。被动语态是科技类文章、产品说明书中常见的语态。其固定结构是系动词 be+ 动词的过去分词。

② Can you tell me the reason for doing sth. 是常用的向他人询问的句式。类似的用法还有 I wonder..., I'm keen to know..., Do you happen to know...? I hope you don't mind my asking, but...。

③ so as to 后加动词原形表示"为了，以便"，和 in order to 的意思相同，但是 so as to 不能用在句首。例如, She got up early so as to/in order to catch the first bus. 她早早起床以便赶上头班车。

④ It's amazing! 常用于对别人进行赞美，尤其是别人完成了很困难的任务或完成任务又快又好超乎想象。

⑤ detail 是细节的意思，复数形式是 details。例如, No details of the discussions have been given. 会谈的细节尚未透露。in detail 是固定短语，表示详细地，具体地。例如 Then we show you how each method works in detail. 然后我们将通过例子，向您详细说明每种方法的使用方法。

Activity 3: Word puzzle. Solve the Cross-word Puzzle with the words of 6S. 字谜游戏。用 6S 词汇完成字谜游戏。

Across
3. 素养
4. 整顿
5. 清扫
6. 安全

Down
1. 整理
2. 标准化

Activity 4: Filling in the blanks using the words in the table. 用下表中的词或短语将句子补充完整。

| stand for | so as to | in detail | aim at | introduce |

1. We arrived early at the theatre_____ buy the front seats tickets.
2. The letters "U.S.A."_____ "United States of America".
3. MGM Company_____ a new system for hiring writers.
4. He told us the accident_____.
5. John is now reviewing his maths lessons, _____ passing tomorrow's exam.

Activity 5: Answer questions. 回答问题。

Why is 6S management popular in enterprises?

Dialogue 2　Explanation of 6S Management 6S 管理的含义

W: Sir, I know that 6S **refers to** Sort, Store, Shine, Safety, Standardize, and Sustaining, but what does 6S **indicate**?	指的是 象征
T: Let me explain it one by one.	
W: OK, the first "S" is Sort.	
T: Sort means to **sort out** things. Not only separate necessary things from unnecessary ones, but also **throw away** the useless articles①.	分类 扔掉
W: So, the unneeded items are **discarded**. Then how to understand the second "S"—Store?	丢弃
T: **To begin with**, we need to select a suitable place for **storage**. Then, **label** the name, **location, quantity**, and put it back when finish using the items②.	首先，储存，标注 位置，数量
W: I think Store means orderliness and **arrangement**—A place for everything and everything is in its place.	安排
T: That's true.	
W: How about the third "S"—Shine?	
T: Shine means cleanness—to keep your working environment clean.	
W: Oh, I see. To clean the working area and make it shine! Safety must be the most important one, I guess③.	
T: Do remember that safety means both safe environment and staff **safety awareness**④.	安全意识
W: Sir, how can we **implement** the fifth S—Standardize?	实施
T: Make **regulations** and every employee do everything according to them.	规定
W: Does the last "S"—sustaining mean good habits?	
T: **Strictly speaking**, the key lies in **self-discipline**, employees **eliminate** bad habits and cultivate good ones to keep the other 5S alive with **ongoing improvements**.	严格来说，自律，消除 持续改进

Note:

① Not only separate necessary things from unnecessary ones, but also throw away the useless articles. 不仅要区分有用和无用的物品,并且要把没用的物品扔掉。Not only ... but also ... 是表示递进关系的连接词,意思是"不仅……而且……"。

② To begin with, we need to select a suitable place for storage. Then, label the name, location, quantity, and put it back when finish using the items. 首先,选择合适的存放地点,第二步标记名称、位置和数量,并在使用后放回原处。

③ must be 用在肯定句中表示较有把握的推测,只用于肯定句,意为"一定"。例如,Carol must be bored in her job. She does the same thing everyday. 卡罗尔一定烦透了她的工作,她每天都做着同样的事。

④ both ... and ... 是表示并列关系的连接词,意思是"……和……都"。

Activity 6: Complete the sentences according to the Chinese. 根据汉语完成句子。

1. You haven't eaten anything since this morning; you _____.(一定饿了)
2. Since the road is wet, it _____ last night.(一定下雨了)
3. The computer is on, so someone _____ it.(一定在使用)
4. It _____ in the office with no air-conditioning.(一定很热)
5. Look at that house! Those people _____ a lot of money.(一定有)

Activity 7: Which "S" do the following pictures belong to？ 判断下列图片属于 6S 中的哪一项。

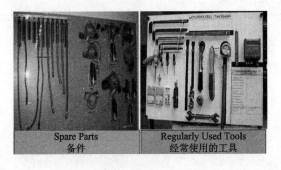
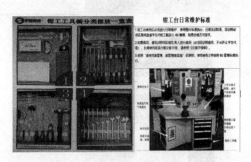

_____   _____

_____   _____

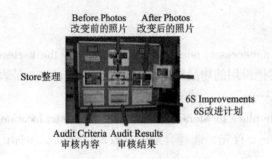

Activity 8: Let's spot the flaws! 大家来找茬!

Context: Suppose you are a team leader and encounter the situations as shown in the following pictures. Please provide suggestions to your team members what he should do.

情境：假如你是班组长，在日常巡检中看到了下图中的情况，请告诉你的组员正确的做法。

1. _____

2. _____

3. _____

Reading and Writing 读写拓展

Reading 阅读

MAHLE (MAHLE Behr Thermal Systems) with its businesses units Engine Systems and Components, Filtration and Engine Peripherals, as well as Thermal Management, ranks among the top three automotive systems suppliers worldwide. MAHLE Behr Thermal Systems (Jinan) Co., Ltd focuses on developing and manufacturing HVAC components and engine cooling systems for vehicles.

The headquarter of MAHLE is located in Stuttgart, which lies in the southwest of Baden-Württemberg in the central Neckar Valley, close to the Black Forest. It is not only the capital of the state, but also the state's largest city. It is also the political center of the state: the Brahman State Council, the state government, and numerous state government departments are located here. Because of its economic, cultural and administrative importance, it is one of the most famous cities in Germany.

thermal [ˈθɜːml] adj. 热的；热量的；保热的　n. 上升的热气流
filtration [filˈtreʃən] n. 过滤；筛选
peripherals [pəˈrɪfərəls] n. 周边设备；[计] 外围设备
cooling system 冷却系统

Activity 9: Read the following passage and finish the questions below. 读短文回答问题。

A study by talent management expert DDI reveals that one in three employees don't consider their boss to be doing an effective job, while nearly half of workers think they could do their boss's job better than them.

Conducted as a way to determine what today's leaders are doing right and what they are doing wrong, the research also finds that fewer than 40 percent of employees are motivated by their supervisor to give their best effort.

Much of the dissatisfaction comes from supervisors' unwillingness to listen to their employees. Thirty-five percent of the surveyed workers say their boss never, or only sometimes, listens to their work-related concerns. Additionally, only 54 percent of supervisors involve employees in making decisions that affect their work.

The study also shows too many leaders aren't delivering on the basic requirements—courtesy, respect, honesty and tact in their interactions—of a healthy manager/employee relationship.

Sixty percent of those surveyed report their boss has damaged their self-esteem, while nearly one-third of employees say their supervisor doesn't remain calm when discussing problems.

The lack of faith in their boss is forcing many to consider other employment. Nearly 40 percent of those surveyed say they leave a job primarily because of their leader, while more than half say their negative perception of a boss has them <u>contemplating</u> finding a new employer.

There are areas, however, where supervisors score high marks. The research shows 74 percent of workers understand their boss's expectations of them, while 66 percent say their manager provides the support they need.

The research is based on surveys of more than 1,200 full-time employees from the United States,

United Kingdom, Australia, Canada, China, Germany, India and southeast Asia.

1. What's the best title for the passage?
   A. The Boss Doesn't Do an Effective Job   B. Hate Your Boss You're not Alone
   C. What's Your Attitude Towards Your Boss D. Is Your Boss Satisfied with You

2. The aim of the study is to determine _____.
   A. what are the right things and wrong things the leaders are doing
   B. whether the employees are giving their best effect
   C. how the boss should motivate the employees
   D. why some of the employees hate the bosses

3. What makes the employees unsatisfied is that _____.
   A. they are not paid regularly   B. the bosses don't pay them well
   C. the working conditions are bad   D. the bosses are unwilling to listen to them

4. Some of the employees changed their job because _____.
   A. they lacked faith in their boss   B. they wanted to get paid more
   C. their boss damaged their self-esteem D. their boss didn't remain calm

5. The underlined word "contemplating" means _____ in this passage.
   A. suggesting   B. imagining
   C. considering   D. managing

Writing 写作

Rationalization Proposal Form 合理化建议表

建议人 Proposer: _____   部门 Department: _____
填表时间 Date: _____

建议名称（主题）Topic of Suggestion:
_____

建议内容（理由）Reasons for suggestion:
1. _____
2. _____

可行性措施和改进方案 Feasible measures and improvement programs:
_____

部门主管意见 Comments from Dept. Manager:
_____

总经理审批 Approval by GM:
_____

Activity 10: Please fill in the table according to the context. 请根据情境完成合理化建议表。

Context：王楠在生产二车间工作期间发现零件摆放架 (Part Rack) 存在占用空间大、难清扫两个问题。建议改造废弃的工具柜 (Tool Cabinet) 用来存放零件。

Part Rack

Tool Cabinet

Letter of Recommendation 建议信

The Letter of Recommendation usually submits the advice to the recipient regarding something in order to let the other party accept his own ideas, claims and solutions.

The Letter of Recommendation can be written to an individual to provide opinions on one problem he/she has encountered; it can also be written to an organization or agency to provide advice on improving its services.

The letter of recommendation should include the reason for the letter, the content of the proposal, and the reasons for the proposal. At the same time, the tone must be appropriate, that is, it must be polite and convincing.

建议信是向收信人就某事提出自己的建议或忠告，以便让对方接受自己的想法、主张并解决有关问题的应用文。

建议信可以写给个人，就其遇到的某个问题提出看法；也可以写给某个组织或者机构，就改进其服务等方面提出建议或忠告。

建议信要给出写信的原因、建议的内容、提出建议的理由。建议信语气一定要得体，既要委婉礼貌，又要有说服力。

The format：

格式：

The first paragraph clarifies the purpose of writing a Letter of Recommendation.

第一段表明写信的目的。

If you write a letter of recommendation to a person, the common opening phrase is usually: I'm glad to receive your e-mail asking for my advice on how to ...

若是给某一个人写建议信，常用的开头语通常是 I'm glad to receive your e-mail asking for my advice on how to ...

If you write a letter of recommendation to an organization, you should first appreciate the other's work, and then euphemistically give suggestions.

若是给某组织或机关写建议信，开头一般应先赞赏对方的工作，然后再委婉地给出想提的建议。

I like it very much mainly for the following two reasons. First, ... Here are a few suggestions. Maybe it is a good idea to ... Would you mind trying ... You should ... // How about ... What about ... //

Why not ... My suggestion is ... // Why don't you ... I suggest ... // I recommend ...

The commonly used ending phrases in the last paragraph are as follows:

最后一段常用的结束语有：

I hope that my suggestions are helpful for your decision-making anyway.

I would be more than happy to see improvement.

I believe that you will take my suggestions into serious account/consideration.

I would be ready to discuss about this matter with you to further details. Whatever you decide to do, good luck with your studies/work!

Dear Jim,

　　I'm glad to receive your letter asking for my advice on how to _____ （引出主题）.

　　Here are a few suggestions. First, it is important to _____. Besides, it should be a good idea to _____. You can also _____. （此三句使用不同的句式提出建议）. As to _____, I suggest _____ （根据需要具体到某一方面）. In addition, _____ （其他建议）. I'm sure _____ （预测可能的结果，给对方以行动的信心和决心）.

　　I'm looking forward to _____ （表达愿望）.

　　　　　　　　　　　　　　　　　　　　　　　　　　　　　　Sincerely yours,
　　　　　　　　　　　　　　　　　　　　　　　　　　　　　　Li Hua

Activity 11: Please fill in the blanks to finish the following the Letter of Recommendation. 请填空完成下面的建议信。

英语老师马丁要求大家对英语课堂教学提出合理化建议。请先简单表达一下你对马丁的课的喜爱，然后对英语课提出建议。

Dear Professor Martin,

　　I would like to take the opportunity to express my appreciation for your humorous and attractive way of teaching. All the information contained in your lectures is comprehensive and practical. To learn more actively, please allow me to outline some of my opinions on your lectures.

　　Firstly, _____ （×××是个好主意）insert a short "intermission" through the lecture and let us have a short discussion. It is helpful for us to avoid decline of attention and catch up with your lecture. At the same time, it takes the pressure off you as the instructor.

　　Secondly, _____ （我建议）you increase student activities or learning tasks during your lecture so as to promote student engagement.

　　_____ （另外）, I hope you will be able to just base your lectures on the most important materials and ask the students to search for the less important information for themselves outside of class.

　　At last, I'd like to express my heartfelt thanks and highest respect to you for your hard work and wish to have an opportunity to communicate with you face to face.

　　　　　　　　　　　　　　　　　　　　　　　　　　　　　　Yours sincerely,
　　　　　　　　　　　　　　　　　　　　　　　　　　　　　　Wang Nan

Activity 12: Please write a Letter of Recommendation according to the context in Activity 10. 请根据练习10的情境写一封建议信。

Speaking Skills 交际技巧

委婉表达

在每一个民族的语言里，或多或少有些关于更礼貌、更客气的表达方法，也就是委婉语气表达。西方人的委婉更是出了名的，英语学习者经过长期的英语学习和研究，不难发现英语中有许多婉转语气的表达法，它使听话者觉得说话者更有涵养或更容易接受其内容，也使自己显得更有教养。那么如何讲有礼貌的英语呢？

委婉提建议

1. I think you should... 我认为你应该做某事，这是很直接的提建议的方式。
I think you should exercise more. 我认为你应该多运动运动。

2. I don't know if... is a good idea. 当你使用这个表达时，你的语气是比较消极的。也就是说，其实你并不觉得这是个好主意。
I don't know if taking a year off work is a good idea.
我不知道休假一年是不是个好主意。

3. Maybe you should try... 和上一个表达不同，当你说"也许你该试着做某事"的时候，语气是积极的，你在鼓励别人去尝试。
Maybe you should try studying a new skill. 也许你应该试着学习一项新技能。

4. I would/wouldn't... would 可以用作劝告，一般用在 I 后面。当你说这句话的时候，你的态度还是比较坚决的。
I wouldn't (= I advise you not to) worry about it, if I were you.
如果我是你，我就不会为它担心。

5. Why don't you...? 你为什么不做某事呢？
Why don't you write out a speech? 你为什么不把讲稿写出来？

6. Have you tried...?/ Have you thought about...? 你试过 / 你想过做某事吗？
Have you tried talking to them? 你试过与他们谈谈吗？
Have you thought about using flowers to apologize? 你想过用花来道歉吗？

委婉拒绝

1. I'm afraid I can't, but ... 恐怕不行……
A: *Will you please come over this Sunday?* 星期天能过来吗？
B: *I'm afraid I can't，but I have some clothes to wash.* 恐怕不行，我要洗衣服。

2. Thank you, but ... 谢谢你，但是……

A: *Will you join us in a walk?* 和我们一起散步，好吗？

B: *Thank you, but I'd rather not. I have something important to do.* 谢谢，可是我不能和你们一起散步，我有重要的事要做。

3. No, thank you / No, thanks. 不用了，谢谢。

A: *Do have another cake.* 再吃一块蛋糕吧。

B: *No, thank you.* 不，谢谢了。

4. I'm sorry, but ... 对不起；很抱歉

I'm sorry, but I must refuse. 很抱歉，我只好拒绝了。

Sorry, but you can't come in. 对不起，你不能进来。

5. I'd like /love to, but ... 我很想，但是……

A: *I hope you can come with us.* 我希望你能同我们一起去。

B: *I'd like /love to, but my mother is ill.* 我很愿意去，但我妈妈病了。

6. I wish I could, but ... 我很希望能……，但是……

A: *The meeting is very important. Can you come?* 这个会议很重要，你能来参加吗？

B: *I wish I could, but I've promised to show Tom around.* 我希望能来，但我已答应带汤姆转转。

7. I can manage. 我能行；我能应付；我可以应付得了。

A: *Do you need any help with these heavy bags?* 你需要帮忙提这些沉重的袋子吗？

B: *No, thanks, I can manage.* 不用了，谢谢，我能行。

8. I'd rather you didn't. 你最好不要……

A: *Do you mind if I smoke here?* 我在这儿抽烟你介意吗？

B: *I'd rather you didn't, actually.* 你最好别抽。

Activity 13: Choose the best answer for the following questions. 请选出下列问句最适合的答案。

1. Will you please come to the meeting tomorrow?

A. No, I have something important to do.

B. I'm afraid I can't, but I have some work to do.

2. We plan to go the beach after dinner. Would you like to join us?

A. I'd love to, but I have to go home early.

B. No, I don't like the beach very much.

3. Would you please give me some advice on English learning?

A. Maybe you should try practicing listening and speaking more.

B. You should try practicing listening and speaking more.

4. Can you give me a favor to clean the table?

A. No, I'm busy now.

B. Sorry, I can't right now, but maybe later.

5. Do you have any plan for weekend?

A. Let's go swimming.   B. Why not go swimming?

Self-evaluation

Items	Yes	No
I have acquired all the key words in the Mission Objective		
I have acquired all the sentence patterns in the Mission Objective		
I can compose a letter of recommendation		

Unit 7

Office English 办公英语

Warming-up 课程准备

Mission Objective 任务目标

1. 掌握复印机、碎纸机、传真机、投影仪等办公设备，期望，进修，总部，深造等英文表达。
2. 会用英语谈论各部门职责、介绍工作职责、表达信心等。
3. 会填写满意度调查表，学习撰写投诉信。

Discussion 讨论

Where do you want to work, in the workshop or in the office?

Words & phrases for assistance:

岗位：post, position, better working conditions, major-related, ability, income, welfare, further development, promotion, work overtime

Database 知识库

Managers in Organizations

Larger organizations generally have three levels of managers: top-level, middle-level, and lower-level managers, which are typically organized in a hierarchical, pyramid structure.

Top

The top or senior layer of management consists of the board of directors (including non-executive directors and executive directors), president, vice-president, CEOs and other members of the C-level executives. Different organizations have various members in their C-suite, which may include a Chief Financial Officer, Chief Technology Officer, and so on. They are responsible for controlling and overseeing the operations of the entire organization. They set a "tone at the top" and develop strategic plans, company policies, and make decisions on the overall direction of the organization.

Helpful skills of top management vary by the type of organization but typically include a broad understanding of competition, world economies, and politics. In addition, the CEO is responsible for implementing and determining (within the board's framework) the broad policies of the organization. Executive management accomplishes the day-to-day details, including: instructions for preparation of department budgets, procedures, schedules; appointment of middle level executives such as department managers; coordination of departments; media and governmental relations; and shareholder communication.

Middle

Middle managers, examples of which would include branch managers, regional managers, department managers, provide direction to front-line managers. Middle managers communicate the strategic goals of senior management to the front-line managers. Middle managers include all levels of management between the front-line level and the top level of the organization.

Their functions include:

- Design and implement effective group and inter-group work and information systems.
- Define and monitor group-level performance indicators.
- Diagnose and resolve problems within and among work groups.
- Design and implement reward systems that support cooperative behavior. They also make decision and share ideas with top managers.

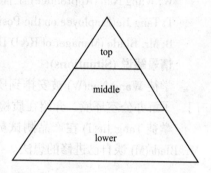

Lower

Lower managers include supervisors, section leaders and team leaders. They focus on controlling and directing regular employees. They are usually responsible for assigning employees' tasks, guiding and supervising employees on day-to-day activities, ensuring the quality and quantity of production and/or service, making recommendations and suggestions to employees on their work, and channeling problems that they cannot resolve to mid-level managers or other administrators. Front-line managers also act as role models for their employees. Some front-line managers may also provide career planning for employees who aim to rise within the organization.

Front-line managers may also provide:

- Training for new employees
- Basic supervision
- Motivation
- Performance feedback and guidance

Activity 1: Fill in the blanks according to the passage. 请根据短文填空。

1. Typical skills of top management includes a broad understanding of _____, _____, _____.

2. Branch managers, regional managers, department managers are _____ managers.

3. Lower managers include _____, _____, _____. They are responsible for assigning employees' _____, guiding and supervising employees on day-to-day activities, ensuring the _____ of production.

Activity 2: Please list all the functions of front-line managers. 请列出一线经理的主要职责。

1. _____ the employees.
2. _____ all the daily works of the division.

3. _____ the workers to improve efficiency.
4. Provide _____ and _____.

✻ Oral Practice 口语训练

人物涉及 (Characters):

J: Mr. Johnson (Quality Inspection Department Office Director) 约翰逊先生，质检部办公室主任

W: Wang Nan (Apprentice) 王楠，学徒

T: Tang Jie(Employee on the Position of Product Test) 唐杰，产品测试员

B: Mr. Blade (Manager of R&D Dept.) 布雷德先生，研发部经理

情景解说 (Situations):

学徒 Wang Nan(W) 被安排到质检部实习，办公室主任 Mr. Johnson (J) 带他熟悉公司各部门，了解办公室环境、介绍在质检部工作期间王楠的工作职责、交代工作任务。

学徒 Tang Jie(T) 在产品测试员的岗位上工作 3 个月了，他在每周面谈时和自己的上司 Mr. Blade(M) 谈自己进修的想法。

Dialogue 1 Arriving at the Office 初到办公室

W: Good morning. My name is Wang Nan, I'm here to see Mr. Johnson.	
J: Hello, I'm Stephan Johnson, the **office director**. Nice to meet you.	办公室主任
W: Pleased to meet you, Mr. Johnson. I'm Wang Nan, from Jinan Vocational College.	
J: Welcome to **Quality Inspection Department**. Let me **show you around**① the **office area** first.	质检部，带某人到处转转 办公区
W: Sure. I've wanted to **take a tour** since I got here.	参观
J: There are many **divisions** here. This is the **Sales Department**, everyone here is responsible for selling products or services.	部门，销售部
W: What about **marketing**?	市场营销
J: **Advertising Department** deals with marketing **affairs**. Let's **move on**.	广告部，事务，继续前进
W: Would you mind telling me what's with the **cubicles**?	格子间
J: This is the **Public Relations Department**, and they are responsible for relating other companies.	公关部
W: I see. The **Human Resources Department** is over there.	人力资源部
J: Yes, a place where hiring② and firing③ happens. This is the **Research and Development Department** which we'll **keep regular contact with**.	研发部 经常联系
W: Ok, I'll **bear** it **in mind**.	记住
J: Now we are at the Quality Inspection Department again.	
W: Shall I meet the **colleagues**?	同事
J: Sure. The guys here are very nice. If there is anything we can do for you, don't **hesitate** to ask. You'll **adapt** yourself to it in **a couple of** weeks.	犹豫，适应，几个
W: I do **appreciate** all you've done for me④.	感谢

Note:

① show sb. around 带某人到处看看。例如：The worker showed us around the toy factory. 一个工人带我们参观玩具工厂。She showed me around and introduced me to everybody. 她带我们到处参观还把我介绍给所有人。

② hire 在句中是雇用的意思。如，We want to hire experienced workers. 我们想要雇用有经验的工人。hire 还有租用的意思。如，They hired a boat to me. 他们租给我一条船。

③ fire 在句中是解雇的意思。如，He was fired by his boss. 他被老板解雇了。

④ 在本句中 do 放在动词前面表示强调。如，We do love our country. 我们十分热爱祖国。在过去时态中，用 did 表示。如，I did see him last night. 我昨晚的确看见他了。

Activity 3: Please translate the following words into English. 请翻译下列词语。

1. 广告部 _____
2. 公关部 _____
3. 人力资源部 _____
4. 销售部 _____
5. 研发部 _____
6. 质检部 _____

Activity 4: Please rewrite the sentences using the word "do" to show emphasis. 请用"do"将下列句子改写成强调句。

1. I care about it.
   _____.

2. I saw him last night.
   _____.

3. Be careful when you cross the street.
   _____.

4. Some people believe that nuclear power poses a threat to the world peace.
   _____.

5. To my great joy, the plant looked exactly like what we were looking for.
   _____.

Dialogue 2　Getting to know the Equipments and Duties 了解办公设备和工作职责

J: Wang Nan, this is your cubicle, desk, arm chair, a small bookshelf, a computer, a telephone and a **cabinet** for storing files. The **photocopier**, **fax machine** and **shredder** are over there.	柜子，复印机，传真机 碎纸机
W: I noticed that in FESTO, the shredders are as **common** as **water coolers** and filing cabinets.①	常见，饮水机
J: Totally.② You do need to deal with the documents according to their **security classification**.	安全级别
W: OK, I'll strictly follow your instructions.	

J: Besides, you are responsible for answering telephone calls and preparing for meetings. We hold department meetings every Friday morning, and there are project-based meetings **frequently**.	频繁地
W: Then could I have a look at the **conference room**?	会议室
J: OK. Here you can see the **projector**, PC, **laser pointer**, whiteboard and marker. Now, let's check if the projector is set up③. By the way, you have to check it for every single meeting, and that's why I suggest you come here at least 10 minutes before the meeting.	投影仪，激光笔
W: It seems that I have a lot to learn.④ I'll try my best.⑤	
J: Hope you have a pleasant stay in the short period of internship.	
W: I'm confident that I can **accomplish** my duties successfully.	完成

Note：

① as...as 意为"和……一样"，表示同级的比较。其基本结构为：as+ adj./ adv. +as。例如，This film is as interesting as that one. 这部电影和那部电影一样有趣。其否定式为 not as/so + adj./ adv. +as。例如，This dictionary is not as/so useful as you think. 这本字典不如你想象的那样有用。as...as 的常见句型还有 as...as possible 尽可能……Please answer my question as soon as possible. 请尽快回答我的问题。as...as usual/before 像平时/以前一样……She looks as pretty as before. 她看起来和以前一样漂亮。

② totally 的意思是"完全"。totally 是略带夸张的说法，与"yes"比起来，totally 更确信，更投入，给人百分百的感觉。例如，"Did you like the movie？"回答是"Totally." "It's totally worth it." "You should totally go." "It was totally awesome." 如果你问外国友人"Do you like China？"回答也是，"Absolutely," "Totally."。

③ set up 是动词调试（设备）的意思。例如，Setting up the camera can be tricky. 调试相机可能会很费事。其名词形式是 set up。

④ It seems that ... 表示"看起来……"。强调根据一定的事实所得出的一种接近于实际情况的判断，可以说表示事实。例如，It seems that she is happy. 她似乎很高兴。It seems that he likes his new job. 他看起来很喜欢他的新工作。It seems that they don't like the idea. 他们似乎不喜欢这个主意。

⑤ try one's best 要尽力去做某件事情的意思。例如，I'll try my best to help you. 我会尽力帮你。I'll try my best to pass the exam. 我要尽力通过考试，等等。该用法非常口语化，日常生活中经常用到。代表着让别人放心，自己一定会办成某件事情。

Activity 5: Match the name of the office facilities with the right picture. 将下表中的设备名称与对应的图片连线。

A. shredder	B. fax machine	C. printer	D. photocopier	E. plastic-envelop machine

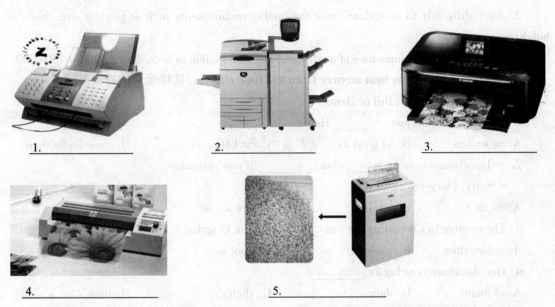

1. _____   2. _____   3. _____

4. _____   5. _____

Activity 6: Filling in the blanks using the words in the table. 用下表中的词或短语填空。

| air conditioning | filing cabinet | paper clip | projector | USB disk |

1. I'll need a _____ to show the trainees the health and safety film.
2. Our office is always cool because of the _____ system.
3. I've run out of staples so you'll have to use a _____ to attach it.
4. I think they are in the _____ under Invoices Paid 2,000.
5. I'll e-mail you the file because my computer hasn't got a _____ drive.

Activity 7: Filling in the blanks using the words in the table. 用下表中的词或短语将句子补充完整。

| Don't phone | dial | Don't e-mail | Delete |
| Don't stop | tick | Don't install | file |

1. _____ the new program until you have removed the old version first.
2. To get an outside line, _____ zero first and then the number.
3. _____ it to her—she hasn't got the right software. Fax it to her instead.
4. Make a copy of the invoice and _____ it in the payments order.
5. _____ the video until they've seen the bit about our new products.
6. Look at the list and _____ the ones you want.
7. _____ him now—he is in a meeting and doesn't want to be disturbed.
8. _____ the old files. Copy them onto a USB disk and keep it safe.

Activity 8: Translate the following Quality Inspector Job Responsibilities. 把下文中的质检员岗位职责翻译成汉语。

Quality Inspector Job Responsibilities

1. Abide by factory discipline and obey the site management of QC;

2. Work diligently in accordance with the quality requirements such as product size, threaded holes test;

3. Perform routine maintenance of quality management system in accordance with the ISO9000.

Activity 9: Choose the best answer from the four choices. 选择恰当的答案。

1. —Who did it better, Bill or Henry?
   —I think Bill did just _____ Henry.
   A. as well as   B. as good as   C. as better as   D. more badly than

2. —The classroom is _____ clean _____ it was yesterday.
   —Sorry, I forgot to clean it.
   A. as; as   B. so; as   C. not so; as   D. more; than

3. The weather in Chongqing isn't as _____ that in Qingdao.
   A. colder than   B. cooler as   C. cool as   D. cooler

4. Our classroom is as big as _____.
   A. of them   B. their   C. theirs   D. their's

Dialogue 3　Face-to-Face Regular Weekly 每周面谈

T: It's my **honor** to have a talk with you, Mr. Blade.①	荣幸
B: The face-to-face **regular** weekly is also a good chance for me to know your **expectations** for future, your ideas to the job or any problems, if you have.②	惯例的 期望
T: I did have a plan to discuss with you, Mr. Blade. You know, I've been on the position of **product test** for 3 months. I **urgently** need theoretical knowledge of **pneumatic circuit**, so I wonder if I can③ **engage in advanced studies**.	产品测试，迫切 气动回路图，进修
B: That's a good idea. **It's never too old to learn.** I'm sure you know exactly what you need now.④	活到老，学到老。
T: The classes are on weekends. I'm full of confidence⑤ that I can **take** both work and learning **into account**.	兼顾
B: Pay attention to language learning. It's **beneficial** in next year's competition to German **headquarter advanced studies**.	有好处的 总部，深造
T: Thanks for your suggestion, Mr. Blade.	

Note：

① It's my honor to do sth. 也可以说 I'm honored to do sth.，"我很荣幸做某事"是口语中常用的客气的说法。还可以更客气地说 It's my great honor to do sth. 例如，It's my great honor to be invited here to give a speech。

② if you have 是条件状语从句，放在句子最后，表示假设，意思是"如果你有"。

③ I wonder if I can 表示委婉的请求。例如，I wonder if I can ask for a leave tomorrow. 不知道我明天能不能请个假？

④ I'm sure you know exactly what you need now. 我相信你非常清楚你需要什么。

⑤ full of confidence 和前一个对话中的 I am confident 都是表达充满信心的说法。例如，Mr. Ryan is confident of success. 瑞安先生对成功充满信心。

Activity 10: Complete the sentences according to the Chinese. 根据汉语完成句子。

1. Please help me open the window, _____. （如果你有时间）
2. You can have a try, _____. （如果你敢）
3. I can give it to you, _____. （如果你想要）
4. _____ introduce the distinguished leaders and guests at this opening ceremony. （我非常荣幸的）
5. _____ use your dictionary for a while. （不知道我能不能）

Reading and Writing 读写拓展

Reading 阅读

VOSS (VOSS Automotive Components (Jinan) Co., Ltd) is invested by VOSS group. As the supplier and expert of commercial vehicle connector technology, it provides product and system solution on pneumatic brake quick connector, SCR line, fuel line and thermo management.

VOSS GROUP is headquartered in Wipperfürth, which is a municipality in the Oberbergischer Kreis of North Rhine-Westphalia, Germany, about 40 kilometers north-east of Cologne, and the oldest town in the Bergischen Land.

automotive components　汽车零部件
commercial vehicle　商用车
connector technology　连接器技术
pneumatic brake　气制动，气压制动器
quick connector　快速接头，快速连接器
fuel line　燃料管线
thermo [ˈθɜːmoʊ] adj.　热的；热电的

Activity 11: Read the following passage and finish the questions below. 读短文回答问题。

When I started my career in management, I was really green, but I always tried to stay positive on the job. I was working for a new company that was having a difficult time with its customers. Actually, at one point my General Manager told me he liked me and suggested I should start looking for a new job as we were about to lose our only order. Every day we faced people leaving the company, customers complaining, and upper management preparing to close the place.

One night, I went home thinking about the problems. My wife told me to forget about work and see a movie. We went and saw Slum dog Millionaire. Jamal, an eighteen-year-old Indian young man, is from a poor family and becomes rich later. He tries his best to change his position. I realized what we needed to do was to correct the situation.

The next day I called my customers to have a talk. I promised to do my best to solve the problems if they would give us some time. As a result, we were given one month to turn the situation around. This was our only chance to correct the situation. To achieve this, I worked day and night with my team. I can remember working from 6 a.m. and not going home until 2 a.m. the next day.

To make a long story short, we succeeded in solving all of the problems and at the same time increased our productivity. The customers and management were very happy, and I knew at that moment we had saved our jobs. Actually things went so well that we ended up having our pay doubled.

Through this experience I understand that you can achieve what you desire as long as you work at it and stay positive.

1. The General Manager advised the author to look for a new job mainly because _____.
   A. he knew the author didn't want to stay there
   B. his company was too small to hire so many people
   C. he believed the author could get more pay that way
   D. the management was preparing to close the department

2. Who caused the author to decide to change his situation?
   A. The hero in a movie.   B. His wife.
   C. His boss.   D. The customers.

3. How long did the author stay in his department every day during that month?
   A. 8 hours.   B. 15 hours.   C. 20 hours.   D. 24 hours.

4. The experience of the author shows that _____.
   A. if a person does his best and keeps positive, he'll succeed
   B. if a person takes advice from others, he'll make progress
   C. if a person is clever, he'll realize his dream sooner or later
   D. if a person wants to change his fate, he mustn't change his job

5. What would be the best title for this passage?
   A. My work experience   B. My first management job
   C. How to correct your situation D. The importance of working hard

Writing 写作

Customer Satisfaction Survey 顾客满意度调查

Date 日期：_____	
Company name 公司名称：_____	Contact person 联系人：_____
E-mail address 电子邮箱：_____	Fax NO. 传真号码：_____

Dear Sirs:

Thanks for your unremitting support for our company! In order to let us understand more about your requirements and advice, please complete this question form to help us improve ourselves and offer better service! Please fill "√" in the blank that you consider degree of satisfaction, of course you can give your advice and try to finish it within 3 days, kindly return to us by e-mail or fax as below, thank you in advance!

先生们：

感谢您对我们公司的不懈支持！为了让我们更多地了解您的要求和建议，填写此问题表，以帮助我们改进自己，并提供更好的服务！请在您的满意程度对应的空白处填写"√"，当然，您可以提出自己的建议。请在3天内完成并以电子邮件或传真的形式回复我们，谢谢！

Compared with other suppliers, your opinion 与其他供货商相比，您认为：

续表

NO.	Item 项目	Quite satisfied 非常满意		Satisfied 满意		Acceptable 可以接受		Dis-satisfied 不满意		Quite dissatisfied 非常不满意		Reason and advice 原因和建议
		10	9	8	7	6	5	4	3	2	1	
1	Service attitude 服务态度											
2	Quantity of products 产品数量											
3	Disposal of defective products 问题件处理											
4	Improve-ment 改进											
5	Production Capacity 生产能力											
6	Delivery requirements 运输要求											
7	Technical Capacity 技术能力											
8	Marketing 市场营销											
9	Sales service 销售服务											
10	Other 其他											

Total score 总分：
☐ More than 90 points 90 分以上；
☐ Between 90 and 70 points 70~90 分；
☐ Between 70 and 60 points 60~70 分；
☐ Between 60 and 30 points 30~60 分；
☐ Below 30 points 30 分以下

续表

Any complaints 意见：			
Any advice 建议：			
Client 客户名称		Signature 签名	

Letter of Complaint 投诉信

In daily lives, there are often unfortunate things that happen, such as impaired consumer interests, disruption of normal life and work, etc. Sometimes writing a letter of complaint is a good solution.

在日常生活中经常会发生一些不愉快的事情，比如消费时利益受损、正常生活和工作受到干扰，等等。有时写封投诉信不失为一个解决的办法。

The letter of complaint usually includes the following aspects: explain the cause of the complaint and express regret; realistically explain the problem, remember not to exaggerate it; point out the consequences of the problem; make criticisms and dispositions or urge the other party to take measures or propose the desired compensation and remedy.

投诉信通常包括以下几个方面的内容：说明投诉的原因并表示遗憾；实事求是地阐述问题发生的经过，切记不要夸大其词；指出问题引起的后果；提出批评及处理的意见或敦促对方采取措施。或者提出所希望的赔偿以及补救方式。

"Three Steps" of writing Letter of Complaint:

Explain complaints –> Describe the specific situation –> Expected solution

Tips:

When writing a letter of complaint, tone should be calm, solemn, and restrained, and it should not be too polite. However, it is also necessary to stick to the facts and not to carry out personal attacks. At the same time, passive sentences are used as much as possible when describing problems to make the description more objective and reasonable.

写作"三步走"：

说明投诉问题—> 描述具体情况—> 期待解决方案

小窍门：

写投诉信时语气要冷静、郑重、克制，不必过于礼貌。但也要就事论事，不能进行人身攻击。同时，在描述问题时尽量采用被动句式，使描述更加客观合理。

The format of Letter of Complaint is as follows:

英文投诉信格式如下：

Dear _____,

① I am _____（自我介绍）. ② I feel bad to trouble you but I am afraid that I have to make a complaint about _____.

③ The reason for my dissatisfaction is _____（总体介绍）. ④ In the first place, _____（抱怨的第一个方面）. ⑤ In addition, _____（抱怨的第二个方面）. ⑥ Under these circumstances, I find it _____（感觉）to _____（抱怨的方面给你带来的后果）.

⑦ I appreciate it very much if you could _____（提出建议和请求）, preferably _____（进一步的要求）, and I would like to have this matter settled by _____（设定解决事情最后期限）. ⑧ Thank you for your consideration and I will be looking forward to your reply.

Yours sincerely,
Li Ming

Activity 12: Please write a letter of complaint according to the following context. 请根据下面的情境写一封投诉信。

Context：You bought a cell phone in a store last week, and you have found that there is something wrong with it. Write a letter to the store manager to explain the problem, express your complaints and suggest a solution.

_____
_____
_____
_____
_____

❀ Speaking Skills 交际技巧

对话破冰与结尾

一、礼貌开始谈话

在面对外国客户和朋友时，很多人不知道该如何开始交谈。虽说要和陌生人攀谈并不容易，但打开话匣子可能只需要你的一句话。那么应该如何打破尴尬的宁静呢？

1. How are you doing today? 今天过得咋样？

简简单单的一句问候最是贴心，谁会拒绝一个笑脸相迎、嘘寒问暖的人呢？

2. Nice earrings! 耳环不错啊！

甭管是衣服、首饰，还是心情、气色，你都可以借来赞扬一番。

3. Oh, did you hear about... 哦,你听说过……吗?

想挑话头,就要学会抛砖引玉,先来给人家讲个故事。

4. What kind of drink is that? 你喝的是什么?

如果你默认对方是枚吃货,而他又正好在品尝美味,不妨从他嘴边的食物开问,他一定会停不下来。

5. That's a lovely name; are you named after someone?

这名字真不错,是因谁命名的吗?欧美人取名常会借用亲人的名字,在他们翻出七大姑八大姨之前,你可以想想接下来聊点什么。

6. People call me David, but you can call me TONIGHT.

大家都叫我大卫,不过今晚你可以打电话给我哦!你看出这句话的端倪了么?想要没话找话,你还是得懂点儿幽默的。

7. Excuse me, I just thought I should come over and talk to you.

不好意思,我就是觉得该过来找你聊聊。拐弯抹角去搭讪不如一句大实话来得痛快!既然是想过去聊天,就不必忸怩啦!

二、礼貌结束谈话

那么如何礼貌地结束一段对话呢?三步技巧在任何时候都能帮助你礼貌地结束对话。

结束对话,基本上分三个步骤。第一步,说明原因;第二步,提及一下稍后有其他事;第三步,说再见。你可以将这三个部分中的句子相组合。比如,

Look at the time. I'm afraid I have another appointment. Let's get together soon, ok? Bye for now. Take care. 时间不早了。我待会还有个约会。下次再聊,好么?再见。保重。

另一种礼貌的表达方式不是直接告诉对方你很赶时间,而是迂回地跟对方说,

Well, I won't keep you any longer. You're probably busy. Or I'll let you go.

您一定很忙吧,我就不再留你了,或者你可以离开了。

你还可以在这句的后边加上一句,

Let's do lunch sometime or I'll give you a call next week. See you. Bye for now.

下次一起吃饭,或者,我下周打给你。再见。

在稍正式的场合中,比如,会议或鸡尾酒会之类的。你可以说,

It's been a pleasure talking to you. /It's been a pleasure meeting you. I'll call you next week. Or may I call you next week, if you want to? 很高兴与您交谈。很高兴与您见面。我下周打给您,或者,如果您愿意的话,我下周可以打给您么?

在这种非常正式的场合下,你应该省去中间这个部分。你可以说,

*I'll look forward to hearing from you, if that's **relevant**.* 我非常期待与您会面。

It's been a pleasure talking to you. Thank you very much. 很高兴与您交谈,非常感谢。

掌握这三步法则,你可以礼貌地从任何交谈中脱身。

Activity 13: Oral Practice. Suppose it's the first time you report for duty in a company, and you have to get familiar with your workmates as soon as possible, please make a dialogue with your partner according to what we have learned. 口语练习。假设你初到企业报到,需尽快与同事们熟悉,请根据以上所学句型,与同伴合作一段对话。

Self-evaluation

Items	Yes	No
I have acquired all the key words in the Mission Objective		
I have acquired all the sentence patterns in the Mission Objective		
I can compose a letter of complaint		

Unit 8
Lean Production 精益生产

☀ Warming-up 课程准备

Mission Objective 任务目标

1. 掌握精益生产、生产管理、标准化工作、PSQ 管理、质量、生产效率、安全、益处、提升等企业管理中常用词汇的英语表达。

2. 会用英语表达带来益处、委婉地提出建议、与人讨论问题时提出自己的意见和同意别人的观点。

3. 会撰写英文报告。

Discussion 讨论

Why do you think management is important in a company?

Words & phrases for assistance:

管理: management, production, quality, quantity, safety, efficiency, cost, profit, waste, time, space, workflow

Database 知识库

What is Lean Production?

Lean is about doing more with less: less time, inventory, space, labor, and money. "Lean manufacturing", a shorthand for a commitment to eliminating waste, simplifying procedures and speeding up production. The idea is to pull inventory through based on customer demand.

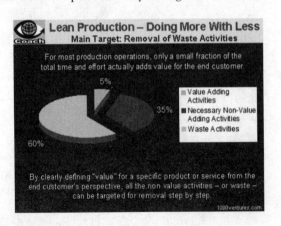

Lean Manufacturing (also known as the Toyota Production System) is, in its most basic form, the systematic elimination of waste—overproduction, waiting, transportation, inventory, motion, over-processing, defective units—and the implementation of the concepts of continuous flow and customer pull.

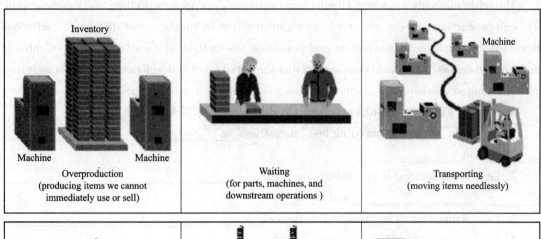

Overproduction (producing items we cannot immediately use or sell)

Waiting (for parts, machines, and downstream operations)

Transporting (moving items needlessly)

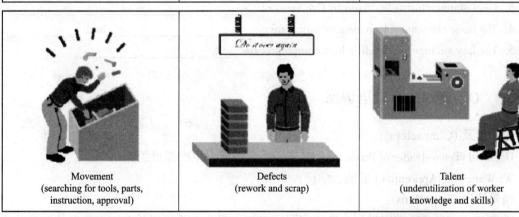

Movement (searching for tools, parts, instruction, approval)

Defects (rework and scrap)

Talent (underutilization of worker knowledge and skills)

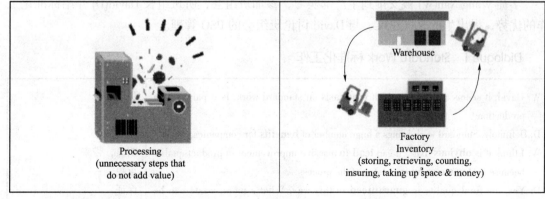

Processing (unnecessary steps that do not add value)

Factory Inventory (storing, retrieving, counting, insuring, taking up space & money)

Five areas drive lean manufacturing/production:
1. cost
2. quality
3. delivery
4. safety

Unit 8 Lean Production 精益生产

5. morale

Just as mass production is recognized as the production system of the 20th century, lean production is viewed as the production system of the 21st century. In fact, the processes involved in lean are ideal for any business whose inventory accumulates in buffer stocks.

The basic elements are waste elimination, continuous one piece workflow, and customer pull. The lean production concept was to a large extent inspired by the Japanese strategy of continuous improvement. Employee empowerment and promotion among them of a way of thinking oriented at improving processes, imitation of customer relationships, fast product development and manufacturing, and collaboration with suppliers are the key strategies of leading lean companies.

Activity 1: Fill in the blanks according to the passage. 根据短文填空。
1. The "less" in "doing more with less" means less _____, _____, _____, _____ and _____.
2. Lean Manufacturing is to eliminate the waste of _____, _____, _____, _____, _____, _____ and _____.
3. Lean manufacturing is driven by five areas of _____.
4. The basic elements of lean manufacturing are _____, _____ and _____.
5. The key strategies of leading lean companies are _____.

✸ Oral Practice 口语训练

人物涉及 (Characters):
D: David (Team-leader of Production Department) 戴维，生产部班组长
W: Wang Nan(Apprentice) 王楠，学徒

情景解说 (Situations):
学徒 Wang Nan(W) 被安排到生产部实习，参加班组会，听班组长 David(D) 介绍标准化工作的优势、提出生产改进建议、与 David 讨论班组会的 PSQ 管理。

Dialogue 1 Standard Work 标准化工作

W: David, it seems that we put great **emphasis** on standard work. Is it part of lean production?	强调
D: Definitely. Standard work brings a huge number of **benefits** for companies.	益处
W: I think it is **obvious** that① it can **lead to** massive improvement in product quality because we **lay stress on** every step of the process.	明显，带来 重视
D: Yes, the final quality is **guaranteed** in this way. What's more, waste can be **reduced** to an **absolute** minimum by following standardized work **principles**.	保证 减少，绝对的，原则
W: From my point of view②, standard work allows apprentices to integrate into **routine** work as easily as possible. We do things according to standard work **procedures**.	日常 步骤
D: **In the long run**, it helps reduce stress and mistakes. So, standard work is a **common practice** used by many companies.	长期 常见做法

Note:

① it is obvious that 中，为了避免主语过长显得头重脚轻，通常在句首用 it 作形式主语，真正的主语是 it can lead to massive improvement in product quality. 句子的意思是：标准化工作可以带来产品质量的大幅度提升是很明显的。除了 obvious 外，easy, difficult, important 等形容词也常用于 it 作形式主语的句子中。例如，It is difficult to get there before dark.

② From my point of view 依我之见，与 in my opinion 可以互相替换。例如，From my point of view, the party was a complete success. 在我看来，聚会非常成功。当然，my 还可以替换成其他的词，例如，From the government's point of view, it would be better if this information were kept secret。

Activity 2: Match the phrases with their Chinese equivalents. 将英语单词与对应的汉语连线。

1. benefit   A. 过程
2. emphasis   B. 步骤
3. improvement   C. 原理
4. significance   D. 提升
5. process   E. 好处
6. principle   F. 重要性
7. procedure   G. 做法
8. practice   H. 强调

Activity 3: Oral Practice. Please list the benefits of standard work according to Dialogue 1. Pay attention to the following verbs and phrases. 口语练习。请根据对话 1 列举标准化工作的益处。注意以下动词和词组。

bring
lead to
is guaranteed
can be reduced
allow sb. to do
it helps

Dialogue 2　Eliminating Wastes 消除浪费

W: David, I heard that we're collecting suggestions on reducing wastes.	
D: Yes. Eliminating waste is the key to improving **enterprise benefit**① . The **final goal** is **elimination** of wastes and zero **losses**.	企业效益，最终消除，损失
W: I noticed some problems during my internship.	
D: You're encouraged to discover problems and give suggestions. If your suggestions are accepted, you'll get **awards**.	奖励
W: Wow, that's wonderful!	
D: The problems that have been found include **over production**, **transportation**, **over processing**, **inventory**, etc. Now, can you tell me what problem you have found?	过度生产，运输过度加工，库存

Unit 8　Lean Production　精益生产

W: I noticed that workers in our team waste lots of time in waiting for parts from the **previous** procedure. Why don't② we **increase** workforce of that team? D: OK, I'll go into **thorough investigations** and if so③, we need to **take measures** in order to make improvements. Good job! W: My pleasure and it's my duty to do that.	先前的，增加 全面的调查，采取措施

Note:

① Eliminating waste is the key to improving enterprise benefit. 消除浪费是提高企业效益的关键。

② Why don't we increase workforce of that team? 为什么不给前一个组增加人手呢？Why don't we... 和 Why not... 一样，是提建议时常用的说法。另外，"I think you should... ""My advice is to..."（"我认为你应该做某事""我的建议是……"）是很直接的提建议的方式。用"Have you tried...?""Maybe you should try..."（"你试过……吗？""也许你该试着做某事"）鼓励别人去尝试。"Have you thought about...?" 多用于比较严肃的场合。

③ if so 是省略了主语和谓语的 if 状语从句，完整形式是 if it is so，意思是"如果是这样的话"。so 在这里指代前文中王楠所说的话。

Activity 4: Filling in the blanks using the words in the table. 用下表中的词或短语将句子补充完整。

investigation	measure	improvement	elimination	award
inventory	production	transportation	procedure	loss

1. Despite thorough _____, the causes of the accident remain unknown.
2. The Panama Canal has played a very important role in _____.
3. The _____ of neck tension can relieve headaches.
4. Industrial _____ is expected to increase by six per cent this year.
5. All of the _____ has been tagged.
6. Your work shows considerable _____.
7. Insurance can protect you against financial _____.
8. The next _____ is to insert the battery.
9. The government has taken _____ to solve the problem.
10. The child smiled at his teacher as he received the _____.

Activity 5: Practising giving advice according to the following contexts. 根据情境练习提建议。

Context: The American supervisor complains about the difficulty in communicating with the local workers and you give him advice on an intensive English training course for the workers.

Clues: listening and speaking ability, oral practice, useful, eager to, communication, learning, attending classes

Dialogue 3 Team Meetings 班组会

W: David, what's the team meeting normally about? D: The team meeting will focus on PSQ which refers to ① **Productivity**, Safety and Quality. W: From my point of view, productivity means we **specify** the **output** ② of the team, for instance, how many products have we finished in this **shift**? Do we meet the **schedule?** ③ D: Absolutely. Safety emphasizes **avoiding** the **incident** or **risky behaviors**, which ④ also enhances safety awareness of employees. W: Quality is the life of an enterprise. D: I can't agree with you more. ⑤ **In terms of** quality, we **inspect** quality problems, more importantly, give solutions to quality problems. W: **Apparently**, team meetings are quite necessary.	生产效率 详细说明，产出 轮班 进度表 避免，事故，有风险行为 在……方面，检查 显然地

Note：

① refer to 的意思是"指的是"，例如，When we say PLC, we refer to the programmable logic controller. 提到PLC，我们指的是可编程逻辑控制器。另外，refer to 还常用于表示"查阅"，例如，Please refer to the table on the following page. 请参考下一页的表格。

② output 在对话中指的是产量，例如，The average output of the factory is 20 cars a day. 该工厂的平均产量是每天20辆汽车。另外，在机械英语中，output 还有输出、输出功率、输出端的意思，例如，output shaft 是输出轴。

③ meet the schedule 的意思是赶上进度。schedule 是日程表、进度表的意思。西方人非常重视日程安排，做事情往往提前安排，因此英语中schedule 是个很常见的词。例如，We are on tight schedule. 我们时间很紧张。behind schedule 延迟，ahead of schedule 提前，on schedule 按进度。

④ 本对话中出现了两个定语从句。大量使用定语从句是英汉表达中的显著区别之一，运用好定语从句能让英语表达更地道。The team meeting will focus on PSQ which refers to Productivity, Safety and Quality. 本句中由which引导的是限制性定语从句，说明PSQ的具体内容，去掉会影响句意。Safety emphasizes avoiding the incident or risky behaviors, which also enhances safety awareness of employees. 这个句子中which引导非限制性定语从句。

⑤ I can't agree with you more. 意思是"我完全同意你的意见。"本对话中"Absolutely."也是同样的用法。另外，我们还可以用 I totally agree with you. /I'm on your side./I think so. /I'm with you.。

Activity 6：Fill in the blanks according to the Chinese. 根据汉语填空并完成句子。

1. Everything went _____. （按计划）
2. The task will be finished _____ if nothing prevents. （提前完成）
3. The airplane reached Paris _____. （按时）
4. My _____ is quite flexible. （日程）
5. Buses are _____ as usual on weekends and public holiday. （安排）

Activity 7: Translate the following sentences. 翻译下列句子。

1. 我认为质量是企业的生命。

2. 更重要的是，班组会能提高员工的安全意识。

3. 我完全同意你的意见。

Activity 8: Oral practice. 口语训练。

Role-play the dialogue with your partner.

Reading and Writing 读写拓展

Reading 阅读

BOSCH (Bosch Automotive Steering (Jinan) Co., Ltd.), a wholly owned subsidiary of Bosch Group, provides the Chinese business partners with a full portfolio of steering system for commercial vehicles. Its hydraulic ball and nut steering gears have been applied by all major Chinese commercial vehicle OEMs.

full portfolio　全系列
steering system　转向系统
commercial vehicles　商用车

Activity 9: Read the following passage and finish the questions below. 读短文回答问题。

How Can SOPs Help with Continuous Improvement Initiatives

Even though standard operating procedures (or SOPs for short) tend to have a poor reputation in certain workplaces, they can actually be a very powerful tool for defining a set of sound rules and guidelines for the way everything in your company is done. If you're striving for continuous improvement, defining a good set of SOPs should be one of the first things you do.

You Need a Consistent Level of Quality

Your company will never truly improve if you don't maintain a good level of quality on a regular basis, and this can be hard to achieve if your operations are not clearly defined. Deviating from the optimal set of steps is one of the main reasons for inconsistent output quality, and preventing that can be as simple as making sure that all rules and regulations of the organization are clearly defined and written down in a concise way.

The Organization Can Become More Flexible

A common misconception about SOPs is that they restrict creativity and make everything monotonous. Some organizations are afraid of implementing them due to the perception that they will make their operations less open to creative input, when the real situation is in fact quite the opposite. Once SOPs are in place, you'll likely find that people are more inclined to share their input on the current state of affairs, and you'll see more suggestions for improvements from all across the board.

Easier Integration of New Employees

Another common problem for companies following continuous improvement practices is that they often run into problems with a large influx of new employees once the organization starts to perform significantly better. This can sometimes be quite problematic, and it's a problem that can significantly undermine the productivity of the company for a long time.

When your work is based on standard operating procedures, new employees will be able to get into their work much more easily, as they will have all their necessary training materials readily available. On the other hand, they will also be able to receive input from their peers more easily, as everyone will be on the same page with regards to how certain things should be done.

Easier Tracking of Faults

And of course, when everything is standardized, finding the root cause of a particular problem tends to get much easier. You won't have to play the dreaded guessing game trying to figure out why something came up, and instead you'll just have to trace over the steps that were carried out and how they deviated from the standard procedure.

Conclusion

Standard operating procedures are one of the best ways to boost the productivity of an organization and ensure that it's always moving in the direction of continuous improvement. When applied correctly, they can completely transform the way processes are carried out, and ensure that your company will never fall off its optimal track at any point.

1. SOPs means _____.
   A. Standard Oriented Performances
   B. State Owned Profits
   C. Support Oriented Processes
   D. Standard Operating Procedures
2. Which of the following statement is NOT the benefit of SOPs?
   A. Maintaining a good level of quality.
   B. Enable organizations to be more creative.
   C. Enable faults easier to be tracked.
   D. Enable employees easier to integrate.
3. What is the main reason for inconsistent output quality?
   A. Deviating from the optimal set of steps.
   B. Rules and regulations defined not clearly.
   C. Lack of standard in manufacturing process.
   D. Lack of quality inspection.
4. If an organization implements SOPs, what will be likely to happen?
   A. They will make their operations less creative.
   B. They restrict creativity and make everything monotonous.
   C. People are more inclined to share their input on the current state of affairs.
   D. You'll see less suggestions for improvements.
5. Why can SOPs make new employees easier to integrate?
   A. New employees will be able to get into their work much more easily.

B. They will have all their necessary training materials readily available.
C. They will be able to receive input from their peers more easily.
D. A, B and C.

Writing 写作

Summary after Production 产后总结

客户 Customer:	订单号 Order No.:
订单数量 Quantity:	物料号 Material No.:
零件号 Part No.:	交货期 Delivery Date:
生产过程问题描述 Description of the Problems during Production Process	
问题原因分析 Analysis to the Problems	

Report 报告

Report is a document used by the subordinate department to report to the higher authorities and leader on certain matters, to report the situation, to make suggestions, etc. In terms of its content, it can be divided into information report and analysis report.

报告是下级部门就某件事情向上级部门和领导汇报工作、反映情况、提出建议等使用的公文。就内容来说，可分为信息类报告和分析类报告。

Information reports are commonly used in everyday affairs. The purpose is to inform information and situations, thus it should be accurate, correct, complete and concise.

信息类报告在日常事务中常用，其目的是通报信息和情况，一般要求准确、正确、完整、简明扼要。

Analysis reports are more formal and complex in terms of language and format. The purpose is to make conclusions after investigating, researching, analyzing, evaluating, or arguing on a problem or situation, which is usually longer.

分析类报告在语言和格式上则更加正式、复杂，其目的在于对某问题或情况进行调查、研究、分析、评估或论证后提出结论，篇幅一般较长。

The writer must first have a thorough understanding of what he has written. When writing, you should use the passive voice as much as possible to avoid being subjective.

撰写人首先要对所写内容有充分的了解。写作时，应尽量多用被动语态，以免主观。

In order to make your report understood easily by the readers, you can use titles, subheadings, numbers, etc, where appropriate. Statistics are often better presented in charts or tables.

为了使读者更容易理解报告的内容，可以在适当的地方使用标题、副标题、编号等。用图表或表格呈现统计数据通常更直观。

The Structure of Report 报告的结构

(1) Briefly describe the background of the report; 简述报告背景；

(2) List the facts or findings; 列出事实或调查结果；

(3) Analyze facts and draw conclusions; 分析事实，得出结论；

(4) Make recommendations based on the conclusions drawn from reasoning, including the specific method of achieving the goal. 依据结论提出建议，包括达到目标的具体办法。

Useful Phrases in Report Writing 报告写作用语

Introduction 简介或事由

- This report aims/sets out to ...
- The report is based on ...
- The aim/purpose of this report is to ...
- This report will summarize ...
- The report provides an introduction to ...
- Upon the request of the committee, we've conducted the report.

Findings 调查结果、事实

- It was found that ...
- The following points summarize our key findings.
- The key findings are outlined below.

Conclusion(s) 结论或推论

- It was decided /agreed /felt that...
- It is clear that ...
- No conclusions were reached regarding ...

Recommendation(s) 建议或推介
- Based on the ..., we'd like to make the following suggestions:
- It is suggested/proposed/recommended that ...
- We strongly recommend that ...
- It is essential to ...
- It would be advised to ...

Example:

Report on Water Supply Facilities

Introduction
The report is based on a check on the water supply facilities on campus in May 2017 to help ease the water shortage problem.

Findings
It was found that most of the hardware works all right but some facilities are problematic. To be exact, 230 out of 1,256 old-fashioned water faucets in school building and student dormitories were found either leaking or broken. Water keeps dripping day and night in 20 student bathrooms for many weeks and at least 12 tons of water is wasted every day.

Conclusions
It is clear that water shortage partially comes from hardware problems and regular check is essential.

Recommendation(s)
I suggest that effective measures be taken to conserve water, including replacing all old water faucets, installing water meter for individual users and enhancing the public awareness of water-saving.

Activity 10: Please write a summary report on production problems and provide suggestions. 请撰写一份关于生产过程中出现的问题的报告并提出改进建议。

```
........................................................................
........................................................................
........................................................................
........................................................................
........................................................................
........................................................................
```

Speaking Skills 交际技巧

定语从句

在中文和英文的表达上最大的区别就是定语。中文习惯把定语放在被修饰成分的前面，而英语恰恰相反，喜欢将定语放在后边，就成了定语从句。

举个简单的例子，如果你习惯说 I like red flowers and green trees, 那么现在就要改成 I like flowers which are red and trees which are green. 值得一提的是，这里并不是建议大家教条地把所有的有定语修饰的句子都改装，只不过提醒大家注意英文的表达习惯，意识到在你的英语口语中应该有大量定语从句的存在，这样才能使你的口语向更为完善、更为地道的方向迈进！

定语从句是由关系代词和关系副词引导的从句，其作用是作定语修饰主句的某个名词性成分，定语从句分为限定性从句和非限定性从句两种。

限定性定语从句

1. that 既可代表事物也可代表人，which 代表事物；它们在从句中作主语或宾语，在从句中作宾语时常可省略关系词。whose 用法相当于 of which。

I like the books which / that were written by Lu Xun.

The desks (which/that) we made last year were very good.

I live in the room whose windows face south. (= I live in the room, the windows of which face south.)

2. which 作宾语时，根据先行词与定语从句之间的语义关系，先行词与 which 之间的介词不能丢。

This is the house in which we lived last year. (= This is the house which/that we lived in last year.)

3. who 和 whom 引导的从句用来修饰人，分别作从句中的主语和宾语，whom 作宾语时，要注意它可以作动词的宾语也可以作介词的宾语。另外，whose, that 引导的从句也可指人。

The old man **who** visited our school is a famous artist.

The old man **whom /that** we visited yesterday is a famous artist.

Miss Wang is taking care of the child **whose** parents have gone to Beijing.

The man **with whom** my father shook hands just now is our headmaster. =The man **whom** that my father shook hands with just now is our headmaster.

4. where 是关系副词，用来表示地点的定语从句。

非限定性定语从句

限定性定语从句是先行词不可缺少的部分，去掉它主句意思往往不明确；非限定性定语从句是先行词的附加说明，去掉了也不会影响主句的意思，它与主句之间通常用逗号分开，若将非限定性定语从句放在句子中间，其前后都需要用逗号隔开。

关系代词 which 在非限定性定语从句中所指代和修饰的可以是主句中的名词、形容词、短语、其他从句或整个主句，在从句中作主语、动词宾语、介词宾语或表语。

1. which 指代主句中的名词。如：

These apple trees, **which I planted three years ago**, have not borne any fruit.

这些苹果树是我三年前栽的，还没结过果实。

Water, **which is a clear liquid**, has many uses.

水是一种清澈的液体，有许多用途。

2. which 指代整个主句。如：

In the presence of so many people he was a little tense, **which was understandable**.

在那么多人面前他有点紧张，这是可以理解的。

When deeply absorbed in work, **which he often was**, he would forget all about eating and sleeping.

他经常聚精会神地工作，这时他会废寝忘食。

Activity 11: Multiple Choice. 选择正确的答案。

1. The freezing point is the temperature _____ water changes into ice.
   A. at which   B. on that   C. in which   D. of what

2. He failed in the examination, _____ made his father very angry.
   A. which   B. it   C. that   D. what

3. Is this the factory _____ he worked ten years ago?
   A. that   B. where   C. which   D. the one

4. Didn't you see the man _____ ?
   A. I nodded just now   B. whom I nodded just now
   C. I nodded to him just now   D. I nodded to just now

5. Can you lend me the novel _____ the other day?
   A. that you talked   B. you talked about it
   C. which you talked with   D. you talked about

6. May the fourth is the day _____ we Chinese people will never forget.
   A. which   B. when   C. on which   D. about which

7. George Orwell, _____ was Eric Arthur, wrote many political novels and essays.
   A. the real name   B. what his real name
   C. his real name   D. whose real name

8. There are two buildings, _____ stands nearly a hundred feet high.
   A. the larger   B. the larger of them
   C. the larger one that   D. the larger of which

9. There were dirty marks on her trousers _____ she had wiped her hands.
   A. where   B. which   C. when   D. that

10. Jim passed the driving test, _____ surprised everybody in the office.
    A. which   B. that   C. this   D. it

📖 Self-evaluation

Items	Yes	No
I have acquired all the key words in the Mission Objective		
I have acquired all the sentence patterns in the Mission Objective		
I can compose a Report		

参 考 译 文

Unit 1
Database

亚琛应用技术大学

亚琛应用技术大学，成立于1971年，是德国著名的应用技术大学之一。拥有区域性自然科学教研中心、研究开发中心、技术转化中心。其工程学专业在全德同类大学（应用技术大学，下同）中排名第二位。其不仅有雄厚的科研资金，而且和欧洲最大的科研中心有着密切的学术和科研关系，这个科研中心有4 000多名来自世界各地的科学家和工程师，学生可以在那里做实习和毕业论文。

大学分设12个院系，有近1万名学生，大约有2 000名来自世界各地的国际学生在这里学习，约占在校总人数的21%。教授250名，教职人员450名。这12个院系有30多个专业。课程着重理论和实践的结合，学科设置适应当前应用科学科技的发展，鉴于其教学与科研实力，该校得到联邦政府、州政府、欧盟、各界工商企业的大量资助，吸引着越来越多的学生选择报读该大学。

该大学正朝着国际化办学方向发展，目前已与美、加、英、法、澳等国家的130多所大学建立了合作伙伴关系（包括我国上海医科大学），互相承认学历和转换学分，进入该校读书的学生还可以申请转往各国的伙伴学校。该大学在德国同类大学中开展国际学位课程最多、效果最好。该校面向国际学生开设了使用英语和德语双语授课的课程，学制为4年，毕业后授予学士学位。这些为国际学生开设的课程使用英语授课，专业学习时使用双语教学，使国际学生无需德语基础就可以直接到德国留学。

Reading and Writing

FESTO（费斯托气动有限公司）：全球仿生技术领先企业，气动自动化领域的领导者，在中国气动行业具有举足轻重的地位，其优异的产品质量、出色的问题解决能力和完善的服务获得了业界的一致好评。

FESTO总部位于埃斯林根，是德国巴登－符腾堡州斯图加特行政区埃斯林根县的首府和最大城市，位于斯图加特东南14千米的内卡河河岸，面积46.43平方千米，人口92 299人(2004年)。内卡河从东南方向、西北方向横贯了城市地区，老城位于河的北岸。

Unit 2
Database

机 械 工 程

机械工程与国民经济的发展息息相关，是为各行各业制造并提供机械设备和电气装置的部门，被誉为"国民经济的装备部"。它不仅强调对数学、力学等基础知识的运用，而且将机械结构的设计、加工、制造作为基础，融入自动控制技术、信息技术、计算机科学技术、艺术等，研究和解决在开发、设计、制造、安装、运用和修理各种机械中遇到的理论和实际问题。同时很多行业的发展都离不开机械类行业的技术支撑，例如航空航天、建筑机械、农业机械等。

机械类专业：包括机械设计与制造、机电一体化、数控技术、工业设计、汽车维修等专业。

就业方向：机械工程是工科中的一个大学科。只要使用设备、生产线，都需要机械知识。就业除了教学、营销外，还有生产总监、物流管理、设备管理、质量管理、项目管理、机械产品开发、汽车工业、模具设计制造、CNC工程师等。也可以从事技术含量高的工作（如设备维护、数控维修、环保设备的设计等）。

"机电一体化"是由日本人创造的术语，用来描述机械和电子工程的整合。这个概念似乎不是什么新鲜事物，因为我们可以环顾四周，看到无数机械和电子结合的产品。然而，机电一体化是专门针对产品和制造系统设计的多学科的、集成的方法。它代表了在各种环境——主要是工厂自动化、办公自动化和家庭自动化环境中，执行工作所需的下一代机器、机器人和智能机构。

Reading and Writing

哈斯文克位于德国北莱茵-威斯特法伦州居特斯洛地区。它毗邻埃姆河，位于居特斯洛的西北方向15千米处。德国机械化收割生产商科勒收公司的总部坐落于此城市，科勒收公司为该市提供了众多的就业岗位。

科勒收农业机械（山东）有限责任公司：100多年的德国企业主要从事联合收割机、牧草打捆机、拖拉机等设备的生产。

Unit 3
Database

求职者面试技巧

1. 对雇主，招聘经理和工作机会进行研究

求职面试的成功始于求职者的深入了解。你应该了解雇主、工作要求和面试你的人。你的研究越多，你对雇主的了解就越多，你就能更好地回答面试问题。浏览公司的网站和其他资料并询问关于公司的问题。

2. 回顾常见面试问题并准备您的回应

面试成功的另一个关键是准备回答可能提问的面试问题。首先，向招聘经理询问面试的类型。它是一对一还是一对多；面试官是一个人，还是你会见到该机构的几个成员。试着确定问题，并撰写详细而简洁的回应，重点关注具体的事例和成就。

3. 着装确保成功

搭配一身符合企业文化的着装，努力保证最专业的外观。请记住，即使过于正式也比着装随意好。将配件和首饰保持在最低限度。面试前尽量不要吸烟或吃东西。

4. 按时到达，放松心情，为面试做好准备

面试迟到没有任何借口。努力在预定面试前约 15 分钟到达，以完成填表等工作，并让自己安静下来。提前一点也是观察工作场所动态的一个机会。

面试前一天，整理你的简历并准备额外的副本。最后，请记住包里装几支笔和一张纸方便记录。最后，当你到达办公室时，关掉你的手机。

5. 取得良好的第一印象

面试的基本原则是要礼貌，向所见到的每个人，从停车员到接待员再到招聘经理，致以热烈的问候。雇主往往很好奇求职者如何对待员工？如果你对任何员工粗鲁或傲慢，你的工作机会可能会轻易溜走。当面试的时候，请记住，第一印象，也就是面试官在见到你的头几秒钟就会得到的印象，会决定他要不要继续面试下去。

请记住，在面试的最初阶段，积极的态度和对工作和雇主表达的热情是至关重要的；研究表明，招聘经理在面试的前 20 分钟内会对求职者做出关键决定。

6. 亲自或通过电子邮件感谢面试官

在面试中，礼节性和礼貌性很强；因此，感谢每位面试你的人也就相当重要。在面试开始时和你离开前，感谢每个面试你的人。在面试结束后写封感谢电子邮件或信不一定能使你获得工作机会，但这样做肯定会让你胜过其他没有发感谢信的候选人。

Reading and Writing

STIHL 的总部在魏布林根，是位于德国西南部巴登－符腾堡州的一个城市。魏布林根有人口 52 948 人，面积 42.76 平方千米。隶属于斯图加特行政区雷姆斯－穆尔县，海拔高度 230 米。

STIHL（安德烈斯蒂尔动力工具（青岛）有限公司）是全球油锯和园林动力工具市场的领导者。STIHL 动力锯一直是世界上头号动力锯名牌，地位牢不可撼。

Unit 4
Database

各部门职责

生产部
- 负责公司的生产管理工作
- 按时完成公司下达的各项生产任务
- 分析与解决生产过程中的重大生产技术问题、质量问题
- 指导生产车间做好生产现场的"6S"管理工作

采购部
- 选择供应商
- 准备采购订单
- 管理客户关系并与供应商保持联系

物流部
- 接受/接收和管理销售订单

- 跟踪货物到达的情况
- 传递不合格货物通知书
- 控制采购货物的发票
- 管理库存交付/调度货物

研发部
- 从事新产品开发和管理
- 根据市场情况，制定公司技术战略和发展目标
- 研究行业技术发展趋势，探索新项目和新产品
- 研究了解产品市场动态，提交新产品开发建议书
- 准备新产品开发计划并组织实施
- 试制新产品和市场推广

人事部/人力资源部
- 招募新员工
- 面试申请人
- 培训工作人员
- 解雇员工
- 处理管理和劳动力之间的关系

经理办公室
- 拨打/接听电话
- 处理文件和信件
- 写备忘录和报告
- 文字处理
- 接待客人和访客
- 安排会议
- 预约

客户服务部
- 处理客户的要求并解决问题
- 接受维修并处理客户的投诉
- 及时与客户沟通
- 提供全面和方便的善后服务
- 提供完整的专业售后服务
- 承担安装和测试任务
- 为用户的技术人员提供培训

经理
- 负责公司的主要活动
- 制造和销售
- 制定决策
- 制订计划
- 处理工作中的困难情况

- 处理问题和投诉

Reading and Writing

维尔恩斯多夫是德国锡根维特根斯坦区的一个市，属于北莱茵－威斯特法伦州。南边与罗塔尔山的 kalteiche 峰相连。kalteiche 峰海拔 579 米，不仅是本区的最高点，也是黑塞（拉恩－迪尔）和北莱茵－威斯特法伦州的边界。本市的文化历史犹如"穿越时间的旅行"：从地质学到石器时代和古代文明，到中世纪和现代。

丝吉利娅提供广泛的门窗相关的产品，在 TITAN, ALU 和 PORTAL 公司的合作下，应用航空和驱动技术，成为世界领先的通风和建筑技术的硬件供应商之一。此外，KFV 技术提供全系列现代门锁系统。

Unit 5
Database

5 种办公室常见危险及预防

1. 滑倒和摔倒

滑倒和摔倒可能是由湿地板这种小事引起的。滑倒和摔倒的其他原因还可能包括不平整的地板、工作台表面，杂乱的走廊和工作区。这些可以通过摆放地面湿滑标志、整理公共走廊和人流量大的区域来避免。

2. 火灾

消防安全在每一个工作场所是非常重要的，因为你不希望看到你的工作成果付之一炬！幸运的是，采取很小的预防措施就可以避免火情。检查所有的电源线，更换磨损的电线，避免电源板和插座功率过高，并严格管理取暖装置。此外，确保所有员工都知道灭火器在哪里和安全出口畅通。

3. 视力疲劳

现今绝大部分的工作场所都需要使用电脑，因此视力疲劳十分普遍，眼睛疲劳会导致眼睛干涩和发炎，最终会导致注意力不集中。定时让眼睛离开屏幕休息，保证光线适度都可以预防视力疲劳。

4. 压力

压力是一个最普遍的办公室风险，而且对于工作和家庭生活都有相当大的影响。每个人都有不同的方法来减轻压力，通常的策略包括多休息、避免加班以及有条理地工作。

5. 人体工程学

错误使用办公家具可能导致长期的身体伤害。可能影响你的手腕、脖子和背部。务必让你的办公桌适应你的身材比例。

Reading and Writing

纽伦堡市位于慕尼黑以北 170 千米处，隶属巴伐利亚州、弗莱肯行政区，是德国东南部的一个城市。其是巴伐利亚的第二大城市（仅次于慕尼黑）。截至 2015 年 2 月，人口已达 517 498 人，成为德国的第十四大城市。

代傲表计（济南）有限公司是全球一流的热量表、水表、燃气表、远程抄表系统的供应商。

Writing

亲爱的玛格瑞特、杰克逊及弗朗西斯，

在上周的部门会议上，我们回顾了上个月的绩效表现，并且达成了多项共识。从新人的岗位问题，到即将到来的项目，以及我们可能会面临的潜在挑战。会议纪要现已编写并已校对，各位可以查阅附件查看完整的文字记录。

如需要讨论会议记录中的任何议题，或者发现会议记录与实际会议内容不一致的地方，请与我联系。

谢谢各位关注！

诚挚的
克里斯汀
工程部高级总监秘书

Unit 6
Database

6S 概述

"6S"最初是 5 个以"S"开头的日语单词，后来加入了 safety（安全），形成了 6S。具体地说，6S 指的是整理、整顿、清扫、安全、标准化和素养。

6S 的第一步是整理，是指在工作场所丢弃所有不需要的、不必要的和不相关的材料。在这个阶段，我们去除不需要的物品，只留下每天所需的工具、设备、部件和机器。实施整理这个步骤起到简化任务、有效利用空间的作用，随之而来的是购买物品时谨慎挑选。

6S 的第二步是整顿。我们组织工作区中的所有物品——将所有物品放在指定的地方，以便快速拿取使用再放回到同一地点。根据工作流程使用者仔细选择每个工具、物品或材料存放的正确位置或容器。如果每个人都能快速获得物品或材料，工作流程将变得高效并且工人工作效率会提高。

6S 的第三步是清扫，指的是清洁工作区域使其更美观。清扫一定要全员参与，从操作工到公司老板。通过清扫使环境更整洁是推广公司时一个重要的市场营销手段。

6S 中的安全有两个重点：创造安全的环境和提高员工的安全意识。具体地说，安全就是回顾每个工作过程和检查工作区域，确保没有忽视任何的潜在风险。

标准化是 6S 的第五步。确保有明确的标准，并且每个人都使用最高效的工作方法。正确的工具处于正确的位置，正确的方法和标准以及积极进取的员工队伍意味着工作更高效、浪费更少。

第六步通常称为素养（自律）。我们努力使 6S 成为公司文化的一部分，以确保持续实施和改进。素养是将前 5 个"S"作为一种生活方式进行持续实践的承诺。素养的重点是消除不良习惯和不断练习好习惯。一旦成为素养，员工就能始终自觉遵守清洁和有序的规定，而无需管理层提醒。

6S 的优势在于效率更高、事故更少、质量更高、故障更少。6S 最大的好处之一是工作流程中的问题变得显而易见。

Reading and Writing

MAHLE（马勒贝洱热系统公司）：凭借"发动机系统与零件""滤清系统与发动机外围设备"和"热管理"三大事业部，马勒集团列全球三大汽车系统供应商之一。马勒贝洱热系统（济南）有限公司致力于开发和生产车用空调零部件及冷却模块。

MAHLE 的总部位于斯图加特市。该市地处德国西南部的巴登－符腾堡州中部内卡河谷地，靠近黑森林。它是该州的州首府、第一大城市，同时也是该州的政治中心：州议会、州政府和众多的州政府机关部门均设在这里。由于其在经济、文化和行政方面的重要性，是德国最知名的城市之一。

Unit 7
Database

企 业 经 理

大公司通常有三个级别的管理人员：高级、中级和一线管理人员。这些管理人员通常形成一个一层层的金字塔结构。

高级管理人员

高级管理层由董事会（包括非执行董事和执行董事）、总裁、副总裁、首席执行官和其他 C 级高管成员组成。不同公司的高级管理人员组成也不同，有可能包括首席财务官、首席技术官等。他们负责控制和监督整个公司的运作。他们设定"顶级基调"、制订战略计划、公司政策，并就公司的整体方向作出决定。

最高管理层必备的技能因公司类型而异，但通常包括对竞争、世界经济和政治的广泛理解。此外，首席执行官负责执行和确定公司的整体性策略。执行管理层完成日常的具体任务，包括：编制部门预算、工作流程和时间表；任命中层管理人员，如部门经理；部门间的协调；媒体和政府关系；与股东沟通。

中层管理人员

中层管理人员，包括分公司经理、区域经理、部门经理等。他们为一线经理提供指导。中层管理人员将高级管理层的战略目标传达给一线经理。中层管理人员包括一线层面和顶层管理者之间的各级管理层。

他们的职能包括：
- 设计和实施有效的团队和组间工作以及信息系统
- 定义和监测小组级别的绩效指标
- 诊断和解决工作组内部和之间的问题
- 设计和实施奖励机制，以推动合作；与高层管理者分享想法并做出决定

一线管理人员

一线管理人员包括主管、部门领导和小组领导。他们专注于控制和指导普通员工。他们通常负责分配任务、指导和监督员工的日常活动、确保生产和/或服务的质量和数量、向员工提出工作建议，并将他们无法解决的问题反映给中层经理或其他管理者。一线经理通常是普通员工的榜样和示范。另外，一线经理还为有潜力的员工提供职业规划。

一线经理通常负责：
- 培训新员工
- 基本监督
- 动机激励
- 绩效反馈和指导

Reading and Writing

VOSS（福士汽车零部件（济南）有限公司）是德国福士集团在中国的独资企业。作为商用车连接技术产品和系统供应商，其产品覆盖气制动快插接头、SCR 管路、发动机燃油管路和热管理系统。

VOSS 集团的总部位于维佩菲尔特，是德国北莱茵–威斯特法伦州上巴吉施的一个直辖市，距离科隆东北约 40 千米，是山城地区最古老的城镇。

Unit 8
Database

何为精益生产？

精益就是用更少的时间、库存、空间、劳动力和金钱做更多的事情。精益生产是对消除浪费、简化流程和加速生产的承诺的简称。其理念是根据客户需求拉动库存。

精益生产（也被称为丰田生产系统），其最基本的形式是系统地消除浪费——生产过剩、等待、运输、库存、移动、过度加工、缺陷部件——以及实施持续流动和客户拉动的概念。

推动精益生产的五大领域：
1. 成本
2. 质量
3. 交付
4. 安全
5. 风气

正如大规模生产被认为是 20 世纪的生产体系一样，精益生产也被视为 21 世纪的生产体系。事实上，精益所涉及的过程对于任何库存累积为缓冲库存的企业都是理想的。

其基本要素是消除浪费、连续整体化工作流程和客户拉动。精益生产理念在很大程度上受到了日本的持续改进战略的启发。其中，以改进流程、模仿客户关系、快速产品开发和制造、与供应商合作为导向的思维方式的授权和推广是领先的精益企业的关键战略。

Reading and Writing

BOSCH（博世汽车转向系统（济南）有限公司）是博世集团的全资子公司，主要为其中国商业伙伴提供商用车全套转向系统。公司生产的循环球式液压助力转向机已被国内主要商用车制造商广泛采用。

Writing

供水设施调查报告

介绍
该报告基于 2017 年 5 月对校园供水设施的调查，以帮助缓解缺水问题。

发现
大部分硬件都能正常工作，但有些设施存在问题。确切地说，校园和学生宿舍内 1 256 个老式水龙头中的 230 个被发现有泄漏或破损。20 间学生浴室中昼夜不停滴水数周，每天至少浪费 12 吨水。

结论
很明显，缺水部分来自硬件问题，定期检查是必不可少的。

建议
我建议采取有效措施节约用水，包括更换所有旧水龙头，为个人用户安装水表，提高公众的节水意识。

参考答案

Unit 1

Activity 1

1. University of applied technology
2. technology transformation center
3. engineering
4. internship
5. full-time students
6. department
7. major
8. faculty members
9. theory
10. practice
11. enroll
12. bilingual teaching
13. academic system
14. bachelor's degree
15. graduation

Activity 3

1. CNC Technology, Industrial Process Automation Technology
2. Qingdao
3. dormitory, canteen, supermarket

Activity 4

1. The teaching building and the workshop.
2. Theory and practice.
3. The college and the companies.
4. They are apprentices as well as students.
5. Dual means both.

Activity 5

1. 面对面
2. 背靠背
3. 肩并肩
4. 心有灵犀、意见一致
5. 心连心

6. 并排、一起

7. 手挽手

Activity 6

1-playground 操场　2-canteen 餐厅　3-dormitory 宿舍　4-workshop 车间　5-library 图书馆　6-square 广场　7-campus 校园　8-supermarket 超市

Activity 7

1. helping me wash my clothes
2. giving her a cup of tea
3. looking after
4. giving
5. opening
6. giving

Activity 8

1. multimedia
2. unnecessary
3. disagree
4. enlarge
5. mislead
6. anti-clockwise
7. reactive power
8. anti-corrosion
9. subassembly
10. equipotential

Activity 9

1-C　2-A　3-B　4-C　5-D

Activity 10

Mr./Ms. Jiang

Wang Nan

April. 6th, 2018

Sick Leave

Mr./Ms. Jiang

am terribly sorry

the rain yesterday, I was wet and had temperature at night

sincerely sorry

permission

Activity 11

1-D　2-B　3-C　4-E　5-A

Unit 2

Top 10 university majors:
1. Business administration and management
2. Psychology
3. Nursing
4. Biology
5. Education
6. English language
7. Economics
8. Communication studies
9. Political science
10. Computer and information sciences

Activity 1
1. Industrial process Automation Technology, Industrial Design, Automobile Maintenance Mechanical Design and Manufacture, Mechatronics, Computerized and Numerical Control Technology
2. development, design, manufacture, installation, application, maintenance
3. Japan
4. new, multidiscipline, integrated

Activity 3
1. set down   2. set back   3. set about   4. set up

Activity 4
1. D   2. B   3. C   4. A

Activity 6
1–E   2–A   3–B   4–C   5–D   6–M   7–K   8–I   9–H   10–G
11–J   12–L   13–F

Activity 7
1–B   2–A   3–A   4–A

Activity 8
1. far more important   2. far more enjoyable   3. far more likely
4. far more accurate   5. far more expensive

Activity 10
1–A   2–C   3–D   4–C   5–D

Activity 11
1–E   2–D   3–F   4–A   5–C   6–G   7–B

Activity 12
1. ABCD   2. D   3. C   4. BC

Activity 13

1. It has not yet definitely settled.
2. I absolutely agree with you.
3. We have been particularly / really busy these days.
4. He bought the house which he especially / particularly liked with all his saving.
5. Undoubtedly, he is the pride of China.

Unit 3

Activity 1

1. job interview   2. interviewer   3. employer   4. applicant   5. first impression
6. warm greetings   7. receptionist   8. thank-you e-mails   9. website   10. resume
11. copy   12. job opportunity   13. positive attitude   14. job seeker   15. offer

Activity 3

equipment, government, statement, movement, advertisement

sensitive, productive, expensive, massive, active

Activity 5

Step 1. The companies give recruitment talk.

Step 2. Apprentices make consultations and choose companies they like.

Step 3. Tests: paper test and face-to-face interview.

Activity 6

1-B   2-C   3-A   4-D

Activity 7

1. As long as there is life, there is hope.
2. You can go out, as long as you promise to be back before 11 o'clock.
3. You'll succeed as long as you work hard.
4. As long as you need me, I'll stay.
5. As long as electric current flows through a wire, there is a potential difference.

Activity 9

1. consult with
2. Focus on
3. looking forward to
4. communicate with
5. in advance

Activity 10

1-B   2-C   3-A   4-D   5-F   6-E

Activity 11

1. 吉姆努力工作，期待着他的假期。
2. 弗兰克盼望着晚上的约会。

123

3. 他盼着能去英国。

4. He was looking forward to working with the new manager.

5. We look forward to seeing you next time!

Activity 13

1-A   2-C   3-B   4-A

Activity 14

Apart from   as a result   Most importantly   as well as   Furthermore

Activity 15

1. because/as   2. such as/ like/for instance   3. not only, but also

4. instead of   5. In addition to/Apart from

Unit 4

Activity 1

1. R&D
2. Purchasing
3. Customer Service
4. new technology, new product
5. Human Resources
6. Production

Activity 2

1-E   2-C   3-D   4-F   5-B   6-G   7-A

Activity 3

I'm responsible for looking up dictionary.

I pay attention to filling the blanks.

I focus on translating the sentences.

I prepare/offer the notes.

Activity 4

1. manual machining, including types of tools, cutting tools, inspecting tools
2. knowledge
3. manual machining
4. analysis of simple parts
5. bench drilling machine

Activity 6

1- 活扳手   2- 钳子   3- 锤子   4- 内六方   5- 两用呆扳手   6- 手锯

7- 螺丝刀   8- 锉

Activity 7

1. ①   2. ③   3. ④   4. ②   5. ⑤

Activity 8

1. milling machine   2. grinding machine   3. drilling machine   4. lathe

5. machining center   6. CNC lathe   7. CNC milling machine

Activity 10

apprentice, Assembly Shop, second, Supply Room, worker, he has been working there only for a week, team leader, Workshop, Maintenance Division

Activity 12

1. Global Positioning System
2. Geographic Information Systems

3–A   4–D   5–B

Activity 13

advertisement in the newspaper

employment

secretary of Quality Inspection Department

position

I'm interested in Quality Inspection

I'm good at English

have a try

Activity 14

Dear Mr.Thomas,

I am writing the application for an all-staff training of QC department on Lean Management from March 20th to 21st.

Lean Management is a main stream which is now popular in modern management philosophy. The lecture applied will focus on the principles and methods of lean management. Under the guidance of qualified and experienced instructor, all the staff of QC Department will have a deeper understanding of the concept.

The total cost of it will be RMB 8,000 Yuan. I greatly appreciate any of your favorable consideration of my application.

Yours Sincerely,
Alex Johnson
Quality Inspection Department Office Director

Activity 15

1. Well, you know.
2. Okay, it's a clever question.
3. How should I put it? Let me see... Well,
4. I'm not an expert in this field, but as far as I know,
5. And to be honest,

Unit 5

Activity 1

1. slips and falls, fire hazards, eyestrain, stress, ergonomics   2. fires(flames)
3. work, home life

Activity 2

Hazards	Causes	Preventative Measures
Slips and Falls	wet floor uneven floors and work surfaces cluttered walkways and work spaces	• ensure wet floor signs when needed • de-clutter shared corridors and places of high traffic
Fire Hazards		• check all your power cords • replace any that are fraying • avoid overloading power boards and outlets • monitor the use of space heaters closely • ensure all employees know where fire extinguishers are • make sure emergency exits are clear at all times
Eyestrain	computers are now used in the majority of workplaces for almost everything	• take regular breaks from the screen • ensure the lighting is appropriate for the task at hand
Stress		• take consistent breaks • avoid working overtime • be organized in your work
Ergonomics	incorrect use of office furniture	set up your workstation to suit your proportions

Activity 3

1. very important person   2. United Nations   3. most valuable player
4. information technology   5. central processing unit   6. Computer aided design
6. programmable logic controller   8. Computerized Numerical Control   9. direct current
10. alternating current

Activity 4

1. It is useful for us to learn English well.
2. It is important for him to know this.
3. It is impossible for us to finish the work.
4. It is necessary for us to get to school on time.
5. It is hard for him to pass the exam.

Activity 6

1-B   2-A   3-A   4-D   5-A

Activity 7

1. too...to...   2. took...off   3. put...away   4. for instance

Activity 9

1-R   2-N   3-P   4-G   5-K   6-O   7-M   8-L   9-D   10-J
11-Q   12-H   13-I   14-A   15-F   16-C   17-E   18-B

Activity 11

1–C   2–B   3–A   4–D

Activity 12

1–D   2–D   3–A, D   4–B   5–A

Activity 13

Dear Mr. Black,

    It is great to know that you are coming to visit our company next month. Please give us some further details about your schedule, in particular, your flight number and date of arrival, so that we can make arrangements beforehand.

    Looking forward to seeing you!

Yours truly,
Wang Nan

Activity 14

1. smart, intelligent, wise, bright, knowledgeable
2. large, huge, giant, enormous, gigantic, immense
3. pretty, gorgeous, attractive, elegant, good-looking, appealing, cute, handsome
4. glad, cheerful, delighted, pleased, joyful, merry

Activity 15

1. preserve   2. hit the sack   3. due to   4. is supposed to   5. round the corner

Unit 6

Activity 1

1. sort; store; shine; safety; standardize; sustaining
2. throw away leave
3. organize
4. Shine
5. Standardize
6. safety awareness
7. Sustaining

Activity 2

1. store   2. sort   3. shine   4. standardize   5. sustaining   6. safety

Activity 3

1. SORT   2. STANDARDIZE   3. SUSTAINING   4. STORE   5. SHINE

6. SAFETY

Activity 4

1. so as to   2. stand for   3. introduced   4. in detail   5. aiming at

Activity 6
1. must be hungry   2. must have rained   3. must be using   4. must be hot
5. must have

Activity 7
1. sort and store   2. store and standardize   3. shine and store
4. safety and standardize   5. sustaining and standardize   6. sustaining and standardize

Activity 8
1. The hammer and pliers should be put into their holders.
2. You need to clean the lathe and make it tidy.
3. Remove the vernier caliper, leave only the required equipment.

Activity 9
1–B   2–A   3–D   4–A   5–C

Activity 10

合理化建议表
Rationalization Proposal Form

建议人 Proposer: Wang Nan   部门 Department: Production Department, Workshop II
填表时间 Date: Apr. 6th, 2017

建议名称（主题）Topic of Suggestion: Using Cabinets Instead of Racks to Store Spare Parts	
建议内容（理由）Reasons for suggestion: 1. Parts racks can not make full use of the space. Compared with racks, cabinets can store more parts because of its height. 2. Part racks are dusty and difficult to clean. Part Rack  Part Cabinet 	
可行性措施和改进方案 Feasibility measures and improvement programs: Remold the abandoned tool cabinet, and store spare parts in it.	
部门主管意见 Comments from Dept. Manager: Approved	
总经理审批 Approval by GM: Approved	

Activity 11

It will be a good idea to

it is advisable that

In addition

Activity 12

> Dear Sirs,
>
> I would like to take the opportunity to express my appreciation for your time.
>
> My name is Wang Nan, an apprentice from Jinan Vocational College, and working in Production Department, Workshop II. During the internship, I found that parts racks are not proper holders for storing spare parts. Firstly, parts racks can not make full use of the space. Compared with racks, cabinets can store more parts because of its height. What's worse, part racks are dusty and difficult to clean.
>
> Maybe it's a good idea to remold the abandoned tool cabinet and make it useful again.
>
> I would be more than happy to see improvement.
>
> Yours sincerely,
>
> Wang Nan

Activity 13

1–B   2–A   3–A   4–B   5–B

Unit 7

Activity 1

1. competition, world economies, and politics
2. middle
3. supervisors, section leaders, team leaders, tasks, quality and quantity

Activity 2

1. Provide training for
2. Supervise
3. Motivate
4. feedback, guidance

Activity 3

1. Advertising Department
2. Public Relations Department
3. Human Resources Department
4. Sales Department
5. Research and Development Department
6. Quality Inspection Department

Activity 4

1. I do care about it.

2. I did see him last night.

3. Do be careful when you cross the street.

4. Some people do believe that nuclear power poses a threat to the world peace.

5. To my great joy, the plant did look exactly like what we were looking for.

Activity 5

1–B   2–D   3–C   4–E   5–A

Activity 6

1. projector   2. air conditioning   3. paper clip   4. filing cabinet   5. USB disk

Activity 7

1. Don't install   2. dial   3. Don't e-mail   4. file   5. Don't stop

6. tick   7. Don't phone   8. Delete

Activity 8

质检员工作职责

1. 遵守工厂纪律，服从 QC 的现场管理；

2. 按照产品尺寸、螺纹孔测试等质量要求勤奋工作；

3. 按 ISO9000 质量管理体系要求进行质量管理体系的日常维护。

Activity 9

1–A   2–C   3–C   4–C

Activity 10

1. if you have time   2. if you dare   3. if you want   4. It's my honor to

5. I wonder if I can

Activity 11

1–D   2–A   3–C   4–A   5–B

Activity 12

> Dear Manager,
>
>   I venture to write to complain about the quality of the cell phone I bought last Friday at your store. During the five days the phone has been in my possession, problems have emerged one after another. For one thing, the screen is always black without any reason. For another, the battery is distressing as it supports the phone's operation for only two hours. Therefore, I wish to exchange it for another cell phone or declare a refund. I will appreciate it if my problem receives due attention.
>
> Yours,
Li Ming

Unit 8

Activity 1

1. time, inventory, space, labor, money

2. overproduction, waiting, transportation, inventory, motion, over-processing, defective units

3. cost, quality, delivery, safety and morale

4. waste elimination, continuous one piece workflow, customer pull

5. employee empowerment and promotion among them of a way of thinking oriented at improving processes, imitation of customer relationships, fast product development and manufacturing, and collaboration with suppliers

Activity 2

1–E   2–H   3–D   4–F   5–A   6–C   7–B   8–G

Activity 4

1–investigation   2–transportation   3–elimination   4–production   5–inventory
6–improvement   7–loss   8–procedure   9–measures   10–award

Activity 6

1. according to schedule
2. ahead of schedule
3. on schedule
4. schedule
5. scheduled

Activity 7

1. From my point of view, quality is the life of an enterprise.
2. More importantly, team meetings can enhance safety awareness of employees.
3. I can't agree with you more.

Activity 9

1–D   2–B   3–A   4–C   5–D

Activity 10

Summary after Production
产后总结

客户 Customer: Zhejiang Jianxin Co.Ltd	订单号 Order No.: ORD 201706526
订单数量 Quantity: 4 000	物料号 Material No.: ACT 175285
零件号 Part No.: PPY 170619	交货期 Delivery Date: Oct. 28, 2016
生产过程问题描述 Description of the Problems during Production Process	
1. During the first two weeks, Workshop B5, Unit 3 lagged behind production plan.	
问题原因分析 Analysis to the Problems	
1. Team leader Li Xiaoqing asked for sick leave for 4 days and the production didn't run smoothly. 2. Jiangsu Wantong Company, supplier for blank ACT 175285 delayed transportation because of bad weather.	

Report on Production Problems

Introduction

The report aims to describe the problems during Part PPY 170619 production to help promote future work.

Findings

The key findings are outlined below:

1. Workshop B5, Unit 3 lagged behind production plan during the first two weeks. To be exact, only 86% percent of parts required has been finished producing by the second Sunday.

2. Team leader Li Xiaoqing asked for sick leave for 4 days and the production didn't run smoothly at the very beginning.

3. Jiangsu Wantong Company, supplier for blank ACT 175285 delayed transportation because of bad weather.

Conclusions

It is clear that the lag in production lies in the following two aspects, manpower and raw materials.

Recommendation(s)

Based on the above conclusions, I'd like to make the following suggestions:

1. Choose vice team leaders and cultivate them in advance.
2. Increase raw material storage esp. in winter.

Activity 11

1–A   2–A   3–B   4–D   5–D   6–A   7–D   8–D   9–A   10–A

常用口语句型

Greetings 问候

Speaker A	Speaker B
Good morning/afternoon/evening.	Good morning/afternoon/evening.
How do you do?	How do you do?
Nice/Glad/ Pleased to meet you. Pleased to meet you again.	Nice/Glad/ Pleased to meet you too. The pleasure is mine.
How are you?	Fine, thanks./Pretty well. What about you?
How are you doing? How's everything? How's it going?	Pretty good./Good. Great./ Pretty good./ Everything is going well.

Introductions 介绍

Speaker A	Speaker B
May I introduce myself? I'm A. Let me introduce myself. I'm A. Allow me to introduce myself. I'm A. Permit me to introduce myself. I'm A.	Hi, A. It's nice to meet you.
Say Hi to B. This is B. Tom, let me introduce you to B. I'd like to introduce my son, B. Allow me to introduce B.	I've heard a lot about you. It's great to finally meet you. I've long been looking forward to meeting you.

Appreciation 感谢

Speaker A	Speaker B
Thank you very much. It's very kind of you. Thank you so much for your kind help. Thank you anyway. I do appreciate all you've done for me. Thank you for your attendance.	It's nothing. You are welcome. That's all right. My pleasure.

Asking for and Giving Directions 问路指路

Speaker A	Speaker B
Could you show me the way to the ...? Would you mind telling me the way to the ...? How can I get to the...?	Walk along this road/street and turn left/right. It's about ... meters along on the right/left. Go down this street/road until you reach the 1st/2nd/... traffic lights. Turn right/left. At the end of the road/street you'll see the ... on the right/left.

Requests 请求

Speaker A	Speaker B
Would you mind doing... Would you mind not doing...	Certainly not.
Can you tell me the reason for doing ... I wonder if I can... I'm keen to know... Do you happen to know... I hope you don't mind my asking, but... Could you please...	Certainly. Surely. No problem. By all means.

Treat 请客

Speaker A	Speaker B
It's my treat. You are my guest. My treat.	Many thanks. Let's go. Thank you, I'd love to.

Advice 建议

Speaker A	Speaker B
Why don't we ... Why not ... I think you should ... My advice is to ... Have you tried ...? Maybe you should try ... Have you thought about ...?	Good idea ! Well, I'll try it. Well, it's a good idea, but... Well, I don't really feel like..., but thank you anyway.

Opinions 观点

Speaker A	Speaker B
From my point of view,... In my opinion...	I totally agree with you. I can't agree with you more. Absolutely. Totally. I think so. I'm on your side. I'm with you on that.

Compliment 称赞

Sounds great!
Sounds good!
Fabulous!
Terrific!
Wonderful!
Fantastic!
Gorgeous!

Responsibility 职责

I'm responsible for...
I prepare...
I offer...
I complete...
I pay attention to...
I focus on...

Opening Remarks 开场白

I made the appointment with you last week for ...
I'm honored to have the opportunity to...

Offering Help 提供帮助

If there is anything we can do for you, don't hesitate to ask.
Let me help you with...

Encouragement 鼓励

Where there is a will there is a way.
Take it easy.

Preference 意愿

I prefer...
I would like...
I'm willing to...

常用前缀后缀

前缀	示例
un-（不，非，表示否定）	unfriendly, unpleasant, uncomfortable
dis-（不，非，表示否定）	disadvantage, dishonest, disagree
bi-（两个，双边的）	bicycle, bilingual
inter-（相互，交互，在一起）	interview, international, internet
re-（又，再，重新）	review, return, rewrite, reset, remove
tele-（远）	telephone, television, teleconference
over-（过分，超过）	overload, overproduction, overtime
micro-（微型的）	Microsoft, microprocessor, micrometer
multi-（多的）	multimedia, multistage
sub-（下面的，下级的）	subway, subtitle, sub-assembly
ex-（向外）	external
anti-（反的）	anti-clockwise

后缀	示例
-or/-er（从事某种职业的人，名词后缀）	visitor, director, singer, runner, worker, operator, inspector
-ist（人，名词后缀）	artist, scientist, tourist
-ese（民族、语言，名词后缀）	Chinese, Japanese
-tion（表示动作、状态，名词后缀）	invitation, attraction, population, production
-ful（充满，形容词后缀）	successful, beautiful, colorful, wonderful
-y（表性质，形容词后缀）	funny, healthy, cloudy, windy
-ing（形容词后缀）	boring, exciting, interesting, outstanding
-ed（形容词后缀）	surprised, balanced, relaxed, talented
-al（……的，形容词后缀）	traditional, international, natural
-able（能够，形容词后缀）	comfortable, unforgettable, programmable
-less（没有，无，形容词后缀）	homeless, helpless, careless, wireless
-ly（副词或形容词后缀）	really, usually, finally, friendly

续表

后缀	示例
–ivity（性质、情况、状态）	productivity, activity
–ment（状态、过程）	treatment, equipment, development
–ware（部件）	hardware，silverware
–wise（方向、方式、状态）	clockwise
–en（使……）	harden

数词的译法

科技英语中大量出现表示数量、倍数增减的词语，如下表所示，翻译时要特别注意英汉两种语言表达方式的异同。

I. 数的单位

汉语	英语
百	hundred
千	thousand
万	ten thousand
十万	hundred thousand
百万	million
千万	ten million
亿	hundred million
十亿	billion
百亿	ten billion
千亿	hundred billion
兆（万亿）	trillion

II. $\left\{\begin{array}{l}\text{as + 形容词 + as + 数词 ...}\\ \text{动词 + as + 副词 + as + 数词 ...}\end{array}\right\}$ 结构的译法

as large as + 数词，可译成"大到……"；
as many as + 数词，可译成"多达……"；
as high as + 数词，可译成"高达……"；
as heavy as + 数词，可译成"重达……"；
as low as + 数词，可译成"低至……"；
以此类推。

The temperature at the sun's center is as high as 10,000,000℃.
太阳中心的温度高达1 000万摄氏度。
The outer portion of the wheel may travel as fast as 600 miles per hour.
轮子外缘的速度可能高达每小时600英里。

III. 倍数的比较

A is N times as large (long, heavy...) as B
A is N times larger (longer, heavier...) than B

A is larger (longer, heavier...) than B by N times

上述三种句型，虽然其结构各不相同，但所表达的概念完全一样。因此，在译成汉语时彼此之间没有区别。三种结构可译成 A 的大小（长度、重量……）是 B 的 N 倍，或 A 比 B 大（长、重……）（N-1）倍。

A yard is three times longer than a foot.
一码的长度是一英尺的 3 倍。

The oxygen atom is 16 times heavier than the hydrogen atom.
氧原子的重量是氢原子的 16 倍。

Mercury weighs more than water by about 14 times.
水银比水重约 13 倍。

This substance reacts three times as fast as the other one.
这一物质的反应速度比另一物质快两倍。

IV. 倍数的增减

increase / reduce N times　增加 N 倍 / 减少 N-1/N

increase / reduce to N times　增加到 N 倍 / 减少到 1/N

increase / reduce by N times　增加了 N 倍 / 减少了 N-1/N

The output of coal has been increased three times as against 1983.
煤产量比 1983 年增长了两倍。

In 1987, the total output value of the textile industry increased to four times compared with 1977.
1987 年纺织工业总产量同 1977 年相比，增长了 3 倍。

A temperature rise of 100℃ increases the conductivity of a semiconductor by 50 times.
温度升高 100℃，半导体的电导率就增加到 50 倍。

The new equipment will reduce the error probability by seven times.
新设备的误差概率将降低七分之六。

The voltage has dropped five times.
电压降低了五分之四。（电压为原来的五分之一。）

V. 分数的增减

分数中，分子部分写作基数词，分母部分写作序数词。如：1/5，读作"one-fifth"。如果分子大于 1，则分母加复数形式，如：3/5，读作"three-fifths"。

The pressure will be reduced to one-fourth of its original value.
压力将减小到原值的四分之一。

The bandwidth was reduced by two-thirds.
频带宽度减小了三分之二。

科技英语中的被动语态及译法

被动语态表示句子的主语与动词表示的动作之间的主动和被动关系。当主语是动作的执行者，即主语是动作实施者时动词用主动语态，如果主语是动作的承受者，动词用被动语态。

科技英语通常真实客观、简洁明确，往往专业性、逻辑性、规范性突出。科技类英语文章中大量使用被动语态，因为被动语态比主动语态更少主观色彩，符合科技英语的特点。另外，被动语态把所要论证、说明的问题放在句子的主语位置上，就更能引起人们的注意。英译汉时，必须注意被动语态的译法，不要过分拘泥于原文的被动结构，而要根据汉语的习惯，灵活多样地进行处理。

科技英语中的被动语态通常用于以下两种情况：

1. 为了突出论述的话题，将所要说明的问题放在句子的主语位置，引起注意。例如，For industrial purposes, materials are divided into engineering materials or non-engineering materials. 就工业效用而言，材料被分为工程材料和非工程材料。

2. 无法指出动作发出者时，直接用被动语态。Various cutting tool materials have been used in the industry for different applications. 多种切削刀具材料被开发出来以满足这些方案中使用材料的多样性。

被动语态的翻译通常有以下几种方法：

1. 译成汉语的被动句，直接翻译成"被……"。例如，

In most facing operations, the work piece is held in a chuck. 在大多数端面车削作业中，工件被固定在卡盘上。

This kind of steel is not corroded by air and water. 这种钢不会被空气和水腐蚀。

汉语表达被动的方式很多，为符合汉语习惯，最好多找一些字眼来代替"被"字，比如可翻译成"由……""用……""受""为……所"等。例如，

The spindle is driven through the gearbox, which is housed within the head stock. 主轴由装在主轴箱内的齿轮箱驱动。

Other processes will be discussed briefly. 其他方法将简单地加以讨论。

Is the air in this area still being polluted by smog? 这个地区的空气还在受烟雾污染吗？

2. 译成汉语的主动句。

1) 保留原文的主语。当英语被动句中的主语为无生命名词，又不出现由 by 引导的行为主体时，常可译为汉语的主动句，原句主语在译文中仍为主语。汉语里这种寓被动意义于主动形式的句子是长期形成的一种表达法，如硬要加上"被"字，则为画蛇添足。

Common surface treatments can be divided into two categories. 表面处理通常分为两类。

If a machine part is not well protected, it will become rusty after a period of time. 如果机器部件

不好好防护，过一段时间就会生锈。

2）把原主语译成宾语，而把行为主体或相当于行为主体的介词宾语译成主语。

Friction can be reduced and the life of the machine prolonged by lubrication. 润滑能减少摩擦，延长机器寿命。

A new kind of magnifying glass is being made in that factory. 那个工厂正在制造一种新的放大镜。

3. 增译逻辑主语。若原文未包含动作的发出者，译成主动句时可从逻辑出发，适当增添不确定的主语，可用泛指的人称代词"人们""我们"等做主语，并把原句的主语译为宾语。例如，

Attentions have been paid to the new measures to prevent corrosion. 我们已经注意到防腐的新措施。

To explore the moon's surface, rockets were launched again and again. 为了探测月球表面，人们一次又一次地发射火箭。

由 it 作形式主语的句型，有时也可以加不确定主语，例如，

Though it has been believed impossible to build a self-generating and ever-running machine, some engineers are trying to develop this idea. 虽然人们认为不可能造出永动机，某些工程人员仍在努力研究这个问题。

英语中这种以 it 作形式主语的被动句型在科技文章中十分普遍，汉译时一般均按主动结构译出，即将原文中的主语从句译在宾语的位置上，而把 it 作形式主语的主句译成一个独立语或分句。现将这类常见句型及其汉译列举如下：

It is generally accepted that... 一般认为，大家公认……

It is believed that... 据信……，人们相信……

It is claimed that... 有人主张，人们要求……

It is declared that... 据宣称，人们宣称……

It cannot be denied that... 不可否认……

It is estimated that... 据估计（推算）……

It has been found that... 已经发现……，实践证明……

It must be kept in mind that... 必须记住……

It is known that... 众所周知……

It is said that... 据说……，一般认为……

It is suggested that... 有人建议……

It is understood that... 人们理解……，不用说……